Flavien Phanzu Mwaka

The field of the universal cosmological model

ISBN numérique : 979-10-262-7511-4

EAN papier : 9791026275121

Librinova

Courriel : contact@librinova.com
Internet : www.librinova.com

LEGAL DEPOSIT/DRC N° : FT 3.02006-57158

From the same author in the collection "The Verses of the Scholars".

Volume I : Cosmology according to biblical and ancient models (French version : ed. Exibook)

First part : Introduction to Pre-Big-bang (éd. Librinova)

Volume II : The field of the Universal Cosmological Model (French and English version : ed.Librinova)

Volume III : Pre-Big bang – Big bang– after Big bang (French and English version : ed.Librinova)

Volume IV : The enigmas of Universe solved in the Universal Cosmological

Model (French version : ed. Librinova)

Volume I : Noah to Barack Obama (French and English version : ed.Librinova)

From the same author in the collection "Astronomy&Theology".

Volume III : Big-bang to humankind (ed. Librinova)

Cover : The shadow of True Parent crisscrossing the cosmos.

N.B. : online sales and in bookshops

Author contact : lesversetsdessavants@gmail.com

Blog : https://payhip.com/LESVERSETSDESSAVANTS

Contents

Epigraph

The eight billion human beings are in fact the eight billion solutions to the problems we face. This is why the development of human beings in every corner of the planet shall be promoted.

Flavien PHANZU MWAKA

Dedication

We dedicate this book to our parents Joseph MPANZU MAKWALA BABAKA and Philomene KIADI UMBA, To my tender wife Djenny MANENE and to our beloved child Nera LEMA PHANZU.

May they find through these dedications, the expression of our love for them.

Preface

Good research work with through undeniable religious, scientific and cultural depth dimension. The covered field topic covered is tough and defensible in the light of reliable material evidence.

This text should be published as an essay balloon for readers unfamiliar with cosmological combined with biblical evidence.

I would like the publication to be partial (3 or 4 volumes) in successive editions and not all at once because of the rigidity of the addressed matter topic addressed.

Once again, thank you and good luck!

Professor GUBARIKA

Dean at the Faculty of Letters
National Pedagogic University (UPN)/DRC.

Foreword

We have divided our books into volumes for the following reasons :

– Volume I constitutes the scientific restitution of the biblical and the ancient creation story. It is the foundation of the Universal Cosmological Model. As the word "Bible" is sacred, I did not want it to be tarnished by criticism of my work or my ideas. So I have detached everything that can be discussed, that is to say everything that can be judged or criticized, I have put it here in this Volume.

– Thus, a judgment such as "I do not agree with the Universal Cosmological Model" is better than hearing someone say "I do not agree with cosmology according to the "Biblical Model". It is to spare the word "Bible" from such criticism that we have detached in the ideas that are part of our contribution to science and put it in this Volume.

– In addition, this Volume is part of my personal enrichment on the study of the universe as the creator of a model. Thus, it proves the place of the Universal Cosmological Model compared to other models. This is to put an end to all those who have labelled us as concordists.

The Universal Cosmological Model field deals with cosmology, not religion. Cosmology is the science that studies the origin and evolution of the universe. We are concentrating for the moment on restoring the first cosmological text discovered on papyrus written by a Jewish scholar in the name of Moses. In order to do this, one must have the knowledge of cosmology and the Principle of Creation of Dr. Sun Myung Moon. These two main disciplines form Astrotheology.

I analyze biblical verses in the same way as I do books and scientific quotations. As an archaeologist of biblical verses, I take what was considered spiritual, I bring it back to earth for scientific analysis.

We should know that that we are not be the last generations of the Universe and our current judgments cannot be taken as absolute. Thirty-five centuries ago, nobody knew that the biblical scheme of creation would pass from religion and cosmogony to science. Today it is analyzed in cosmology through its scientific restitution in the Universal Cosmological Model.

In spite of my research and knowledge, I have not reached a full understanding of this scientific restitution and I am not yet able to express it perfectly. This consideration concerns especially the pre-Big-bang, because the Post-Big-bang is widely analyzed by science even though it made a bad start by making the Big-bang the point of origin of the Universe therefore excluding the biblical scheme of creation.

What I present in the books of the series "The Verses of the Scientists", are only introductions to open the door to the masters of this discipline who are astrophysicists, cosmologists, astronomers, planetologists, physicists, astrochemists, theologians and all those who devote themselves to the study of the origin of creation, in order to allow them to study and develop it in a scientific way for some and in a theological way for others.

However, no one will really be able to discover the keys to the Universe in a perfect way. Even though we have within us the spirit of the Christ the one who witnessed these things (Jn 1:2), we are only the result. Besides, in science it is said that an answer brings up other questions. As long as we have not yet reached the Cause, we will always have questions.

When God told me that all models should be based on the biblical scheme of creation, even if no one can believe me as the relationship with God is personal from one individual to another, I shuddered at the thought of writing it in my book, because at that time I didn't know how to defend it. But the real defense will only come from the scientists themselves, who were only waiting for the biblical scheme of creation to be clarified, even though the terms are sometimes unsuitable for science. The real terms for science are the results of research. Therefore, scientifically the Universal Cosmological Model is the result of the research I do, since I bear full responsibility for it. But we cannot dismiss God as the hero in the shadows.

The knowledge of the teachings of the Divine Principle of Rev. Dr. Sun Myung Moon and the preliminary notions in cosmologies, allowed me not only to clarify the biblical scheme of creation but also the Ancient Models and exploited by the various sects, including esoteric sects. It also allowed me to offer a clear explanation of the creation of man according to the Koran and to address certain hypotheses in cosmology which science will decide on later.

So, there are scientists in astrophysics and exegetes of the word of God, but before the appearance of this book, with what Saint Paul calls having optional gifts (Rm 12:6), I was the only one in the universe to make a connection between these two disciplines and to discover that the history of creation according to the Bible is scientifically logical, richer and more coherent than all the models that science has presented to us.

By the way the "Universal Cosmological Model" swallows up and sweeps away the other cosmological models, while integrating them, correcting them and putting each in its context. This is the subject of this volume.

(Working session at the Observatory of Baronny Provencal (OBP) in France / September 2019).

Behind me, Mr. Marc Breton ; the director of the Observatory and our Master (September 2019)

Introduction

Some readers will read this volume without having had the opportunity to read the previous ones. Therefore, we will go back to the basics of the Universal Cosmological Model for a better understanding of this volume.

Moses, through the Bible, was the first to speak of the pre-Big-bang to man according to what this book has just revealed to humankind. Before, nobody knew that the first day of Genesis in the Bible was about the pre-Big Bang or Before Big-bang. This is also because nobody knew that the Big-bang was biblical.

The Post-Big-bang period has been widely exploited by science, it is now up to us to take a look at the Pre-Big-bang period by trying to review the models of all those, who in the past, had to think about it. These models come in support of the Biblical Model, clarified in the Universal Cosmological Model. Therefore, all the research on this period before the Big Bang we call it the Field of Moses.

Believers from the astrophysicists or atheists, no one has a choice because Moses is the first to be able to offer a Model and he is the first scholar in this field. The fact that he divided the scheme of creation according to the days proves the introduction of mathematics.

In fact, all the Models to this day present themselves as five blind people who were told to touch an elephant and each one keeps to themselves the true image of the part they palpated as an elephant. The one who touched the tail will say that the elephant looks like a rope, he is not wrong, except that his description is not complete. Such are the Cosmological Models that have been found so far in science.

The universal cosmological model presents the elephant in its entirety. And God was right to reproach me for wanting to equate the models of today's scientists with the biblical scheme of creation. Then, he insisted that everyone should follow the biblical model, logically restored in the Universal Cosmological Model, and it is up to the scientists to enrich it.

In the books of the "Verses of Scholars" series, we are developing an old model for religion, but a new model for science : The Universal Cosmological Model, which has just swallowed up the other cosmological models that have held the attention of scholars for over a century.

The scientists who were called "prophets" in the past were very wise and intelligent people who reasoned beyond what others thought were limitations. Nowadays, they are people who advance science either by their thoughts, theories or discoveries, based on research.

Some astrophysicists will consider the Universal Cosmological Model to be an unscientific model because it is beyond the scope of the latter's work. Who gave limits to the scientific work? A limitation of the knowledge in physics does not mean that we have reached the end. What I analyze is also part of the scientific work. It is part of Astrotheology or the combination of theology and science.

Up to now, contemporary physics has not given the answer to all the observations and questions that it asks itself. Indeed, some of the laws it has brought to light do not coincide in the fields where they are supposed to act jointly or concurrently. Among those that are identified, we have :

- why the pre-big-bang sphere is still considered a mystery?
- Why doesn't quantum physics match relativistic physics?
- Where do masses come from?
- Why is there matter and not so much antimatter?
- Why is the expansion of galaxies accelerating?
- etc...

These questions have not been fully resolved in any of the proposed models. Physics adds ad hoc additions to its models such as inflation theory, symmetry breaking, the Higgs boson, and others. In spite of these additions, and even if some of them were to be verified by observation, it would still be necessary to have a theory that would allow the calculation of masses, the confirmation of the dates and modes of formation of matter, light, stars, galaxies and the acceleration of the expansion of the Universe, etc. This is by no means the case in current mathematical models.

According to Faber Sperber, they all have the disadvantage of using a mathematics of numbers and a continuous geometry, as well as a non-dialectical logic (Cf : matierevolution.fr/spip.php? auteurs2Dark matter, dark energy : the missing link? 13/11/2007, by Faber Sperber, Robert Paris).

First of all, the cosmos is a divine projection. Even if mathematics can explain things, it only acts on the level of effects. A theory that everyone expects to shed light on certain mysteries level must be able to bring to the surface the cause that is contained in the spiritual. Therefore, it is useless to expect that it comes from an astrophysicist, who, by the way, is already limited by classical theories. On the other hand, only astrophysicists can enrich it, because they are the ones who hold the key to observations. And here I only repeat what God told me, namely : "Everyone must be inspired by the biblical creation scheme". This is the only original model that everyone should follow and enrich. What has been developed since Volume I of this book to open the way for astronomers is called "The Universal Cosmological Model".

Finally, as in the previous volume, I do not use the classic system of the "Footnote Method" and only use the "Author-Date Method in the Text".

Chapter I :
The field of the Universal Cosmological Model

We define the field in the Universal Cosmological Model as the field of study covered by this model.

Section 1
The Biblical Model and the Universal Cosmological Model

1.1. The Biblical Model

The Biblical Model is the story of creation as told in the Bible in the first two chapters of Genesis. The restitution that was made in Volume I in a scientific way is called the Universal Cosmological Model.

The biblical scheme of creation had never attracted the attention of scientists and was rarely the subject of sermons. Until this book appeared in bookshops, there was no one to develop its theme. The reason is very simple ; this story belongs to the scientific and religious field and to discover it, in addition to biblical knowledge, one needs knowledge of the preliminary notions of cosmology and the Principle of Creation contained in the book of the Divine Principle by Rev. Dr. Sun Myung Moon. This is the basis of Astrotheology.

As the first person in humankind to have forced himself to learn all three knowledge at once until the appearance of this book, it is understandable why I am the first to do the symbiosis. That is to say that understanding the story of Genesis about the creation of the Universe in a scientific way was not as easy as one might think, except after reading this book.

Thus, branches such as, the preliminary notions of cosmology, astronomy, the anatomy of living beings, the Divine Principle of the Moon couple, and the biblical scheme on creation scientifically based on the Universal Cosmological Model must be included in all theological, philosophical and science studies. Without this inclusion, the philosophical thinking issued by the latter will always remain partial.

These disciplines are part of Astrotheology. These are branches that allow a theologian or philosopher to have a say in science. Otherwise, he risks engaging in a debate without understanding what he is saying, as Saint Thomas Aquinas said "*Also, when a truth is scientifically demonstrable, he who knows it only as an*

opinion, for a reason that is only plausible, does not understand it".
(Saint Thomas d'Aquin)

1.2. The Universal Cosmological Model

The Universal Cosmological Model, we have mentioned, is the analytical restitution of the biblical scheme of creation. Analytics is defined in philosophy as a conclusion from principles based on experience and observation (Dicos Encarta).

It is the starting point from which everyone can research and draw comments and conclusions. That is to say that no one, not even I, has the monopoly of analyzing the Biblical Model.

The latter is the set of all the Models that support the biblical pattern on the creation of the Universe. I have said that what is contained in the series "Astronomy and theology" is called "the Universal Cosmological Model" which everyone can learn and enrich from.

My only progress in relation to the analyses that will come later to support the biblical scheme of creation is to have opened the door and to have it scientifically approved, even if the arguments are recorded in a simple reading book. This is the only way I can enter everywhere and teach the humankind.

My great contribution is also to have clearly explained the words, names and sentences that made the biblical scheme of creation seem like a magician's game with phrases like "God said, God made, God created".

1.2.1 The Universal Cosmological Model statement

Thus, the Universal Cosmological Model on which all the subsequent analyses of the various scholars and philosophers will be based on is as follows :

"The Universe was formless and empty. Given, on the one hand, an energy defined as darkness by the Bible in the abyss, and on the other hand, the spirit of God or a wind from God on the original material called "waters", there was the production of a radiation defined as light. From the beginning of the Universe in nothingness until the first light or radiation was produced, this is the period that

the Bible calls 'day' (first day), a period that lasted for a time that could not be quantified until then.

The Bible further specified that this new energy or radiation had an effect on darkness. It came to counteract it, as well as the invisible material that was already at work in the early universe. This is expressed in the sentence "God separated the light from the darkness". And to show that these two energies would henceforth accompany the physical destiny of the Universe, God called darkness "night" and light or new radiation "day", for lack of the appropriate scientific words at that time.

The original material or "waters" according to the biblical term, continuing its warming cycle after the production of light (light be and light was), had a burst, an expansion or an expulsion into the abyss and that is what science calls "Big-bang" and the Bible calls "separation of the waters from the waters. This evolutionary period of material warming (waters) from the production of light or first radiation, to the Big-bang, was described by the Bible as the second day. And then came the third day and the fourth day which marked the beginning of the post-Big Bang period which were periods of the creation of the stars from the same material or waters according to the Bible.

Then came the fifth biblical day, the period that witnessed the appearance of living beings from microscopic underwater beings to birds.

And at the end or on the sixth biblical day, there was the appearance of mammals and of humans. (Gen 1:1-30; Gen 2:1-7).

Since Moses, the editor of Genesis, later referred to the day after the six days as a weekly Sabbath, many believers and non-believers think that each day of creation must have been of 24 hours.

We explained the concepts of "day, night, morning and evening" extensively in Volume I. In spite of the biblical verse : "But one thing, beloved, which you must not ignore, is that before the Lord a day is as a thousand years, and a thousand years are as a day" (2 Pet. 3:8), no Christian theologian has been able to relate this verse to the days in the biblical scheme of creation, believing that the latter was the work of a word. It took a new revelation of God to the Moon couple for the enlightenment to stand on the basis of this verse.

Rev. Dr. Sun Myung Moon is the first religious man to have associated them with the biblical scheme of creation in the Principle of Creation; a branch of the book of Divine Principle. Let us take a look at the verses again :

Ps 90 : 4 " *For in your sight a thousand years are like yesterday that passes by, like a few hours of the night. »*

2 Pet 3:8 "But one thing, beloved, which you must not ignore is that with the Lord a day is as a thousand years, and a thousand years are as a day. »

There are two very important words in these verses, namely "to ignore" and "as". First of all, we should not ignore the notion of time. Secondly, there is the comparative word "as" : "a day for God is like a thousand years". Here, the word "equal" has not been used, but rather "as". This simply means that a biblical day can represent as many days, as many years as a cycle or an era can take.

So to say that creation was made in six biblical days has nothing to do with our twenty-four hour days. Rather, it represents six great stages which, as science has discovered, took thousands, millions, even billions of years.

Moreover, the biblical account does not exclude the possibility that during each "day", or period of creation, some of the major creative acts occurred gradually rather than instantaneously. Hence the words "evening and morning" were used to move from one period to the next. For more details, I refer you to Volume I, the explanations taken from the book "Presentation of the Divine Principle". *(Cf : Presentation of the Divine Principle : ed. Culture et Paix, Paris, 2013).*

Finally, it should be noted that in Genesis, Moses wrote that the Universe had a beginning and that things and life appeared in stages, gradually over several periods. How did Moses gain access to such precise scientific information? It was only the Creator of all these works, God our Father, who revealed it to him as it is written :

Ex 33:11 "The LORD spoke with Moses face to face, as a man speaks to his friend".

Section 2
The Universal Cosmological Model field and the Moses field

2.1. The Universal Cosmological Model field

The field of the Universal Cosmological Model is the expansion of the field of study it covers. We have said that the Standard Model studies of macroscopic physics while quantum mechanics studies the microscopic physics, the result does not allow it to be included in the Standard Model. Each Model proposed by astrophysicists has its field of study.

The Universal Cosmological Model constitutes the scientific restitution of the biblical creation scheme. Its field covers and studies the events from the moment the word Universe was first uttered in the Bible through its definition of a *"formless and empty earth*" (Ge1:2), that is to say, something that had no material existence yet, otherwise it would never be formless and empty, until the creation of humans. This study is limited to the level of origins and also attempts to unravel the mystery of "why".

The Universal Cosmological Model sweeps the areas before the Big-bang, from a formless and empty earth, called the quantum vacuum, through the Big-bang to man. What other Model could impose it ?

It only remains for astrophysicists, who are the specialists of this science, to offer a plan from the spare parts, to review all the Models and to glue each piece back in its place. And, in the end, this will be the culmination of the Universal Cosmological Model.

2.2. Field of Moses

2.2.1. Field of action

Ge 1:1 "In the beginning God created the heaven and the earth. "

Gen 1:2 The earth was formless and empty : there was darkness over the surface of the deep, and the spirit of God moved over the waters. 3 And God said, let there be light. And the light was. 4 And God saw that the light was good; and God separated the light from the darkness. 5 And God called the light day, and the darkness he called night. And there was evening and there was morning : and it was the first day.

We have developed in detail all these pre-Big Bang structures in Volume I which is entitled "Cosmology according to biblical and ancient models".

We said that the field of Moses covers all the research on the period Pre Big-bang. He is the first scholar who actually spoke about it and whose writings are contained in the Bible. This is what made it a spiritual scheme and science did not want to take into consideration. But, through its restitution in the Universal Cosmological Model, the veil has just been lifted and we find ourselves in the presence of a scientific scheme.

Chapter II
Pre-Big-bang or before Big-bang

We offer a short summary for those who have not yet read previous volumes. The Universe experiences three great periods according to the Universal Cosmological Model : the Before-Big-bang the Big-bang and the after-Big-bang.

Section 1
The meaning of the word Big-bang

1.1. Definition

The Big-bang represents the zero moment when the present Universe began to exist. It was the blossoming of the condensed Universe described by the Bible as "the separation of the waters from the waters". The Pre-Big-bang period is a period that is not yet scientifically recognized. It is in the biblical scheme of creation written 1500 years BC by Moses and the writings we discovered of the philosopher Empedocles who lived in the 5th century B.C. While for science time and space it did not exist during that period, but through the Bible we discover that it was a very eventful period, described as the first day, as we can read :

Ge 1:2-5 "The earth was formless and empty, and darkness was over the surface of the deep, and the Spirit of God was moving over the waters. 3 And God said, Let there be light. And there was light. 4 And God saw that the light was good; and God separated the light from the darkness. 5 And God called the light day, and the darkness he called night. And there was evening and there was morning : one day.

Empedocles described the events that took place Pre– Big-bang as follows : "*Originally, all the elements, united by Love, formed a homogeneous mass called the "spherus".* (Cfr : Empedocles, Wikipedia).

Here, we are talking about the elements that were united, before a homogenous mass called the "spherus", effectively demonstrating that there was a force that he called "love" and that it was also an eventful period. We have extensively developed the Pre-Big-bang period in the first part of Volume I. Here it is just a little reminder.

The field of MOSES covers the period before the Big-bang until the appearance of the radiation emitted by the first biblical light and its great implications in the Universe through the biblical phrase : "...God separated the light from the darkness...". It is not in philosophical or theological thinking that we can interpret this sentence, but it was physical phenomena taking place in the Universe before the Big-bang or "the separation of the waters from the waters". This is why the development or research on the basis of the biblical scheme of creation is a matter of cosmology.

It is this light that will reveal much more about darkness or what we have called dark energy and dark material. Perhaps science has already detected this light, because for me, as I said, it is fossil radiation. Why do we first suspect fossil radiation as the biblical light before Big-bang ? We have discussed this in details at Volume I, even if another dimension of explanation might be necessary.

1.2 Origin of the word Big-bang

This name, which was previously known only to scientists, is becoming more and more famous in the general literature. This name has even come to create a total separation between those who thought that it was God who created the universe and science where the majority of scientists brandish it as being the physical creation of the universe without the help of a Creator.

First of all, what does the Big-bang mean ? The Big-bang represents the instant zero of our universe. Indeed, according to the accepted theories, our universe started as an atom, let's say as a grain of sand for those who do not have notions in chemistry, and experienced a fulgurating expansion, called Big-bang.

This word was pronounced for the first time by the British scientist Fred Hoyle in 1950 who said that the universe has no beginning and will have no end. Then to mock the astrophysicists who said that the universe began as a grain of sand and experienced an expansion, he asked them the question : "You mean that the universe has experienced a Big-bang ? Since then this word has remained famous until it was cast as a valid theory about the beginning of the universe.

Observations and scientific evidence only confirm that, our universe did start as a grain of sand and experienced a meteoric expansion, qualified to this day as a Big-bang.

If this is part of the beginning of the universe, then why does the biblical creation diagram, which is the first cosmological diagram to describe the beginning and the evolution of the universe, not mention it? I am the first researcher to be able to look into this question and what I discovered surprised me and everyone else.

The biblical creation scheme contains many mysteries. If I have managed to unravel some of them, it is due to my elementary knowledge of cosmology and the Principle of Creation of Dr. Sun Myung Moon.

To deny to this day that there was no Big Bang is like denying that air does not exist. The Big Bang did happen and is indeed part of the evolution of the universe. The cycle of the universe knows three great steps and it is me who affirms it. It is about :

– The Pre-Big Bang; characterized by a contraction of energies until reaching a point of minimal concentration;

– This point of minimal concentration constitutes the second step called the condensed universe resembling a primitive atom or a grain of sand. No one can give the dimension with exactitude. The Big-bang is associated with this second step because it was a transit step between the pre-Big-bang and the post-Big-bang.

– This atom has been expanding until now, the beginning of which was called the Big-bang. And, this expansion constitutes the third step called the after Big-bang.

In my research, I discovered that this pattern, which is recognized by all scientists today, rather confirms the Bible, whose creation story was written 1500 years ago before Jesus Christ. Let's. read how the scholar of scholars Moses presented it in the Bible :

Gen 1:2 "The earth was invisible and empty (Universe) : there was darkness on the surface of the deep and the spirit (a wind) of God moving over the waters. (Septuagint Version)

Ge 1 : 6-8 "God said, 'Let there be an expanse between the waters, and let it separate the waters from the waters.' 7 And God made the expanse, and separated the waters that were under the expanse from the waters that were over the expanse. And it was so.8 God called the expanse heaven. So there was evening and there was morning : this was the second day

Section 2
The origin of the universe according to the Bible

Most scientific models have offered an answer as to how the universe began. These answers are divergent from each other. But the universe came into being only once ; this means that there must also be only one answer. All these divergent answers are the fruits of the classical models. However, no one has paid attention to the answer of the first book that offered a cosmological scheme, namely, the biblical creation scheme. It is only attributed to creationism, which all, religious or scientific, have confused with apparitism. To create a thing or a being does not mean to make it appear. All creations require work and even evolution is included in it before the created being reaches maturity or final form. So what does the Bible say about the creation of the universe. Everyone's answer will be "In the beginning God created the heavens and the earth" (Gen 1:1). The second question is how did God create the universe ? For the Vatican or the Holy Seat, it is to science to offer the answer. But for the Universal Cosmological Model, the Bible interprets itself and it is up to it to offer the answer.

2.1. The beginning of the Universe according to the Bible

Gen 1:1 In the beginning God created the heaven and the earth. 2 Now the earth was invisible and empty; darkness was over the deep, and the Spirit of God was upon the waters. (Septuagint)

Gen 1 : 1 In the beginning God created the heavens and the earth. 2 The earth was formless and empty ; darkness was over the surface of the deep, and the Spirit of God was moving over the waters. (Louis Segond)

Gen 1:1 In the beginning God created the heaven and the earth. 2 The earth was without form and void, and darkness covered the waters ; the active power of God moved over the waters. (New world)

Gen 1:1 In the beginning God created the heaven and the earth. 2 Now the earth was empty and waste, and darkness covered the deep, and the wind of God moved over the waters. (Jerusalem Bible)

If in theology such a change of words in bold or of sentences has no importance, but in cosmology it has heavy consequences. Now when we say "In the beginning God created the heaven and the earth", we are in theology. But when we address the problem of the beginning of the earth in verse 2, we enter into cosmology, and there each sentence differs from the other according to the different versions. How can we create a unity of language with science if within religion everyone has his own version?

While science is confined to the Big Bang, during more than a year of my research, I could not understand the point of view of the Bible on the beginning of the Universe. I was doing my analysis from the 3 versions that the religion presents to us, namely Louis Segond of the protestants, Jerusalem of the Catholics and New World of the Jehovah's Witnesses. Their differences made my task difficult because they are the ones who translated the Bible from the original language into everyday language. Finally, I fell back on the Louis Segond version when I discovered that a formless and empty earth simply means that materially the earth had no existence or was simply invisible. This is what I have developed in my work, concluding that according to the Bible the universe was initially empty, that is, without any structure inside. It is in this emptiness that the structures appeared on the third and fourth biblical days, and the living beings from the fifth biblical day.

I could even say that this point of view on the beginning of the universe was scientifically confirmed by the Italian scientist Gabriele Veneziano in the Cosmological Model of Strings without him even being aware that he confirmed the Bible.

While my synthesis according to which the universe began in a vacuum was criticized by many believers claiming to be creationist fixists, i.e. that God created things as we see them today, and that on the side of science too, I was accused of concordism, I have just discovered a biblical version which has just given me reason. This biblical version is missing from the popularizations and it was completely unknown to me until I discovered it by chance on a certain 25/08/2021, that is to say after the publication of my books or the results of my researches. It is the Septuagint Bible or LXX; translated from Hebrew into Greek in 270 years B.C. by 72 Jews including 6 translators per tribe in 72 days. Hence the name of the Septuagint version. It was translated into French in 1870 by P.GIGUET. (Cfr : https://fr.wikipedia.org/wiki/Septante)

In the Septuagint Bible, we read this concerning the first 2 verses of Genesis : "In the beginning God created the heaven and the earth. 2 But the earth was invisible and empty...". So, which scientist or atheist will tell me again that I made a concordism when I said that "a formless and empty earth" means that materially the universe was invisible at its beginning because it was empty? And, what do we see in the void when there is nothing? We see nothing except a space that extends to infinity. And, the "one" also represents nothing, except a fictitious witness.

This is what the Italian scientist Gabriele Veneziano has confirmed scientifically by saying that the Universe comes to us from an infinite void. Certainly for the believers, the infinite is God, but that does not change anything geographically neither for them, nor for the science on the nature of the universe has its origin. It is this Septuagint version that was torpedoed by other translators who did not understand what could mean "an invisible and empty earth" because of the word earth. So they used their own reasoning instead of faithfully translating what was written. The word universe did not exist in 1500 BC, so Moses simply defined it as "an invisible and empty earth" to make a difference with our planet "Earth" which was created on the third biblical day.

And, at the beginning I said that in the biblical creation scheme, the biblical days are periods that can go from one millisecond to more than one billion years. In fact, 2P3:8 says "Do you not know that to God a day is like a thousand years. Like a thousand years, not equal to a thousand years, so it can vary. It is all about the time of the fulfillment of an ideal. It is this fulfillment that is qualified as a day, no matter how long it took, before we move on to another ideal or to the other day.

2.1.1. Difference between the heavens and the formless and empty earth

Ge 1:1-2 *"In the beginning God created the heavens and the earth. The earth was formless and empty.*

In Volume I, we said that at the very beginning of the Universe, there was only a void as the Bible says "the earth (Universe) was formless and empty". But what is to say that we were not in front of two spheres where we could distinguish between the spiritual sphere (the heavens) and the material sphere (the Universe), as they were for a fictitious observer visibly empty? And how is this emptiness filled today with physical structures?

The answers come to us from science and this through quantum physics. Talking about a formless and empty earth, therefore materially non-existent, physicists have found that the void is filled with physical energies which can be virtual or elementary. Virtual because they are impossible to detect. This is the answer to the first question that allowed us to distinguish between the heavens and the earth. In the heavens we find particles that cannot be physically detected. Now, through the radiation or the divine projection, a sphere was formed where these particles became dense in relation to the spiritual elements, becoming the so-called virtual particles, contained in a formless and empty earth or the Universe at its beginning.

For the French physicist and philosopher Etienne Klein, "the vacuum is not nothingness, but rather a kind of prologue of matter : if we give it energy, it is capable of producing particles that in fact it already contained, but in a latent form, virtual, only potential. (Cfr : Klein, Etienne, les secrets de la matière, éd. Plon, Paris 2008, p.66).

Through this quotation namely by giving energy to the particles contained in the vacuum, Moses in the Bible mentioned it by the verse "A wind of God (energy) moved over the waters (particles)" in the pre-Big-bang era.

The second question was : How is this void filled today with physical structures? This question can also be formulated as :

"How can we imagine the beginning of the Universe ? As I said, the answer comes to us from science in quantum mechanics, but we preferred the development of the chemical physicist Ilya Prigogine who wrote in "Time to become" :

"One essential aspect is the transition from virtual particles to real particles. Virtual particles are particles that are created in pairs, that live for a certain time and then recombine as a result of Heisenberg's uncertainty principle. The real particles, on the contrary, do not have to recombine anymore; they are not held by a thread to each other. So the creation of the Universe, the creation of the matter as we know it, corresponds above all to a creation of possibilities, creating at the same time disordered phenomena and highly organized phenomena". (Cfr : http://www.matierevolution.fr/spip.php? article38The vacuum, ... not so empty, the23/06/2010, by Robert Paris).

The vacuum is thus full of virtual particles, then we understand that the real is nothing else than a transformation exerted on these virtual particles.

Thus, to this day, no astrophysicist can contradict me when I say that quantum physics is part of the physics of creation and consequently, it is part of Before Big-bang. For the believers, in this quote, the word 'disorderly' is not appropriate in case it contributes to the parameters that make life possible on a part of the Universe according to the anthropic principle.

In conclusion, we can say that at the very beginning of the cosmos, according to the biblical language, the heavens represented the spiritual sphere and the earth, the physical sphere or universe. If the spiritual sphere is and remains invisible, the physical sphere was rather empty. The difference was that the spiritual sphere or heavens was made up of spiritual particles, while the physical sphere (universe) or a formless, empty earth was made up of virtual particles.

2.2. Difference between the biblical heaven and the geographical heaven

Gen 1:1 *"In the beginning God created the heaven and the earth"* (Jerusalem Version)

Gen 1:6 *God said, "Let there be an expanse between the waters, and let it separate the waters from the waters." 7 And God made the expanse, and separated the waters that were under the expanse from the waters that were over the expanse. And so it was. 8 God called the expanse heaven.*

The meaning of the word "heaven" seems to be commonplace, yet each person has an incomplete view of the word. On April 12, 1961, as part of the Soviet space program, Yuri Alekseevich Gagarin, a Soviet pilot and cosmonaut, after a flight in space during the Vostok 1 mission, said a sentence that pleased scientists : "I have travelled through the heaven, but I have not seen God. And yet, Christians pray to "Our Father (God) who is in heaven". Is this the same heaven ?

To clarify this question, we have called upon our basic knowledge of cosmology and the Creation Principle of Dr. Sun Myung Moon. The word heaven is first found in the first sentence of the Bible where we read "In the beginning God created the heaven and the earth".

In Volume I, I said that the heavens referred to in the verse "In the beginning God created the heavens and the earth" is the spiritual world, just as "the earth" in the same verse represents the physical world, defined by "a formless and empty earth", that is, the universe. Therefore, "God created the heavens and the earth" according to the Bible means "God created the spiritual world and the physical world. It is in these heavens or skies, unreal for science, that all religions locate the invisible spiritual entities : God, angels, wandering spirits, paradise, hell.

But still in the biblical creation scheme, if we go down a little bit (Gen 1:6-8) we are told about "the separation of the waters from the waters" or Big-bang according to scientific terms. After the Big-bang or at the Big-bang, the universe had entered an expansive step called by the Bible extensive. And the Bible says, it is this expanse that we call heaven. There, now, we are in front of the physical heaven. The scientists themselves confirm this by talking about the beginning of space-time. It is different from the spiritual heaven which is mentioned in the first verse : "In the beginning God created the heaven".

The physical heaven which is mentioned in the eighth verse,

"God called the expanse heaven" is the geographical heaven. It is the empty space visible to our senses. This void that separates me from the other, that allows me to move, this void that contains everything in the universe. The real term of the physical heaven is the void or the space. If we combine them, it becomes the space vacuum. Thus, Gagarin did not fly in the heaven, but rather in the vacuum of space. And, in the vacuum of space, we can see neither God, nor angels, nor wandering spirits.

So, let science leave the words "heaven or skies" to the religious. Religious people must also understand that the words heaven and skies have nothing to do with the void of space, nor with the blue heaven that we see above our heads. It is rather the spiritual sphere that our physical senses cannot see or feel.

In conclusion, when the Bible and all the sacred books speak of "the heavens or heaven", they refer to the spiritual world, unreal for science. When the general literature and science speak about the heaven, they refer to the spatial void or geographical heaven.

Many believers ask me, what do astronomers look for in the heaven ? I answer them : we do not search in the heaven, which is a spiritual field, our search is limited to the space vacuum. Some scientists also ask me : "We search the heaven, but we do not see your gods or your angels. And I answer them : you do not search the heaven because it is part of the spiritual domain. You go through the void of space where we cannot see spiritual beings.

So, in Astrotheology, the heaven and the void of space are two different words. Heaven refers to the spiritual realm, invisible to our senses, while the void of space refers to the physical world perceptible by our senses. Thus, our airplanes and other machines do not fly in the heaven, but in space. To the question of where is the heaven ? It is the spiritual space that contains and encompasses the physical space.

2.3. Gravitational collapse into the void

Gen 1:2 "There was darkness on the surface of the deep".

Gen 1:2 "and the spirit of God moved over the waters". (Louis Segond)

In astronomy, a gravitational collapse is a contraction of either energy, particles or matter under the effect of their own gravitational attraction. Having acquired an overload in space, the body collapses on itself, distorting the fabric of space-time.

After an "invisible and empty earth", we are told that "there was darkness on the surface of the deep" (Gen 1:2). This quote shifts the biblical creation scheme from cosmogony to cosmology. The abyss implies a hole, but why did the scholar Moses use the word "abyss"? According to the biblical glossary "an abyss is a hole whose bottom cannot be seen and where mysterious (unknown) phenomena take place inside and that it is a zone of no return once crossed".

When the scientist John Wheeler in 1967 did research on celestial abysses, he nicknamed the "black holes". And, black holes do meet the definition of abyss. This time it is science that has missed a name for these structures just as the Bible missed the word for the "early universe" and spoke of a "formless and empty earth".

Indeed, the first biblical day is part of the pre-Big Bang era. However, all the scientists who have reasoned about the pre-Big-bang have spoken either of a collapse (abyss) like the Italian scientist Gabriele Veneziano and others of a black hole (abyss) like the British scientist Roger Penrose. Some, like the French physicist J.P. Luminet have hypothesized that the universe would come from a glass hole which is the exit of the black hole. And, the recent studies made at the Perimeter Institute of Canada deny that there was a black hole before the Big-bang.

They have all, in fact, without exception, confirmed the biblical verse "there was darkness on the surface of the abyss", i.e. the existence of a black hole or a mass that dug the space-time. And, the darkness is what also characterizes the surfaces of the black holes. How was created this black hole or this gravitational collapse at the pre-Big-bang? This is what we will see.

2.4. From virtual particles to real particles

Gn 1:2 "*A wind of God whirled on the waters*". (Bible of Jerusalem)

We said according to the Bible that the beginning of the universe was defined as an invisible and empty earth, showing that there was nothing. In the next point, we are told that a wind from God was moving on the waters. How can there be water in a universe that is empty and invisible to a fictitious witness ? By this we understand that even the water spoken of here should also be something invisible.

Indeed, where did the qualified matter of water come from in an invisible and empty universe ? Scientists have discovered that in even the most complete vacuum, there is energy. And, the ancient language to designate energy was spirit. Philosophically, active or positive energy is called Yang, while passive energy is the symbol of negativity or Yin. Scientifically, we speak of virtual particles, forming particles and antiparticles, the whole forming what is called in religion the spirit of God (V. Louis Segond) or a wind of God (V. Jerusalem).

When we say spirit or wind turning or moving, we discover that we are in front of a dynamic phenomenon which the author attributes to the divine power. But it does not change its nature if science calls it gravitational attraction. In a gravitational attraction, we see the bodies approaching or merging, but we do not see the force that makes them approach. This is how Moses called this force a divine spirit or a wind from God, i.e. attributing it to God. But it was a physical phenomenon.

According to CERN experiments at the Laboratory for Nuclear Physics (LHC in Switzerland), the compression of energy creates matter. And, in a formless and empty universe, there was only energy that sometimes manifested itself in the form of virtual particles and that intertwined according to the Uncertainty Principle. Their compression according to the biblical phrase a wind moving on the waters brought them back from virtual particles to real particles.

So, the biblical sentence that described what happened in an invisible and empty earth is the following : "a wind or a spirit of God moved on the waters". For its scientific development, we referred to the theory of the strings of Gabriele Veneziano which is articulated as follows :

"Originally, the universe was always empty. In this emptiness, gravity eventually took over and the matter (the particles) began to agglomerate. The density of the matter is such in certain regions that it forms an incredibly massive black hole, whose interior knows a process of expansion ". (Cfr : LES SECRESTS DE L'ESPACE " HORS-SERIE " N° 7 ; Les mystères de l'Univers, p.13).

This is how the abyss was formed, to which it is written "there was darkness in the surface of the abyss. And, it is in this abyss that the processes that led the virtual particles to the real particles took place. In conclusion, "a wind of God whirling in the water" was the gravitational attraction that was exerted on the virtual particles, until it culminated on the real particles.

Section 3
The first light

Ge 1 : 3 *God said, Let there be light. And there was light.*

3.1. Celestial radiation

I am introducing this point for my readers to help them understand the different radiations coming from the deep heaven called x-rays and that are not to be confused with the fossil radiation called cosmological diffuse background.

X-rays or cosmic rays or cosmic background X-rays often come from the many stars and galaxies, as well as from regions of the cosmos where there are extreme conditions, like a very hot or invaded region by magnetic fields, as for the black holes. Most of it is absorbed by our atmosphere, which acts as a protective shield. This is why, the more we stand up in space, unprotected, the more we run the risk of being bombarded by these rays that go through our bodies and we don't notice them.

The cosmological background or fossil radiation, on the other hand, is due to the radiation that fills the whole of space and is not associated with a specific body. This is why we attribute it to the first light cited in the Bible through the phrase "*God says let there be light! And there was light* ". (Gen 1:3-4).

3.1.1. Fossil radiation

Ge 1 : 3God says : Let there be light! And there was light. 4 And God saw that the light was good; and God separated the light from the darkness. 5 And God called the light day".

According to several scientific sources, fossil radiation, also called the cosmic background, a radiation that is found in all directions of the heaven. It is interpreted by science as a remnant of the period of intense heat experienced by the Universe 380,000 years after the Big Bang.

The extension of its radiation is now the limit of the observable Universe. It is related to the period when the Universe suddenly became transparent, 380,000 years after the Big-bang. It therefore corresponds exactly to the maximum distance from the Earth to the cosmological horizon, which is the boundary of the observable Universe.

In fact, this radiation, whose temperature is uniform to all galaxies, was considered to be the first proof that the big bang took place and that the original material was initially collected and had benefited from the same temperature. Now, the Bible, by speaking of a wind of God moving over the waters, meant that during material deflation was gathered by what we call the Universal Prime Force, which science speaks of as a super-force. And as a result (I don't often like to talk about the consequences, because for us believers, there was a divine will inside), the elements heated to the extreme until they reached a point where they emitted the light defined by the biblical phrase "*God said that there was light and there was light*" (Ge 1:3) and this was before the Big Bang.

However, when the Universe reached its extreme condensation, even this light could not escape as in a black hole.

After the Big Bang, as I have already said, it for science there followed a period of darkness again, until gravitation began to merge the elements again and light began to appear again 380,000 years later. This is why, by taking a step back from what fossil radiation should have been at its beginning, we are not back to darkness but rather to a mass as we can see in the diagram below.

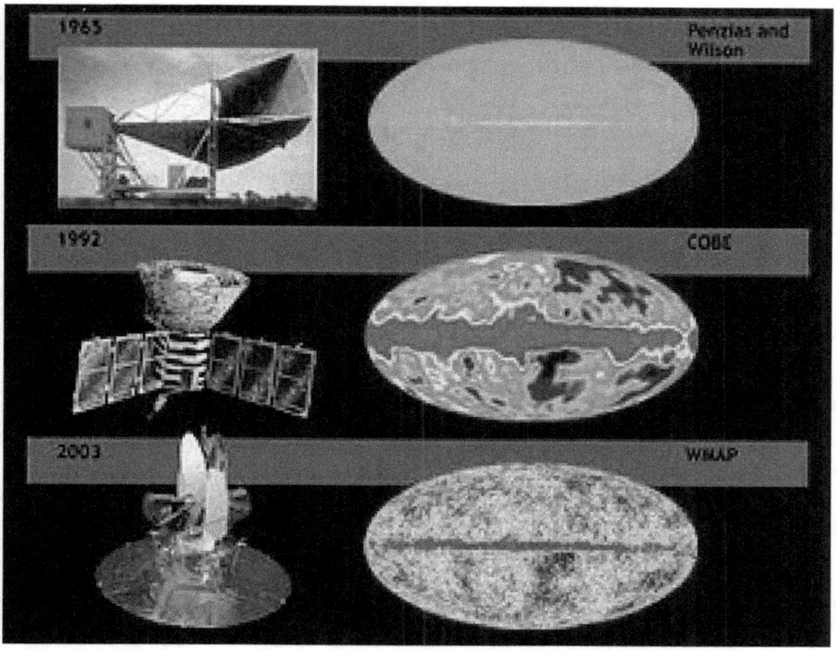

The diffuse cosmological background, discovered in 1965, is the most direct witness of the Big Bang. Since then, its fluctuations have been studied by the space probes COBE (1992), WMAP (2003) and Planck (2009). Detailed article : WIKIPEDIA ; Cosmological background

Let's read some comments of scientists and ours on the interpretation of the cosmic microwave background.

3.1.1.1. Scientific commentary

These images are interpreted scientifically in the following way :

"Operational since 1992, the American satellite COBE (Cosmic Background Explorer), has been mapping the background radiation from the heaven emitted in the microwave range, as shown in this image. Predicted in 1948 by the Russian-born American physicist George Gamow and detected in 1965 by American astrophysicists Arno Penzias and Robert Wilson, the cosmic background is interpreted as a track and proof that the primordial Universe has experienced a very dense and very hot step.

According to the Big-bang theory, this 3 K microwave radiation is a residue of the energy prevailing at the time of the creation of the first stable atoms in the Universe (material-radiation decoupling) – some 300 000 years after the primordial explosion. This observational fact is now one of the pillars of the Big-bang theory, the most likely theoretical model to describe the evolution of the Universe. In the image, the relative temperature variations (red color for the hottest, blue color for the coldest) are interpreted in terms of the density variation of the primordial Universe. These densities in homogeneities are believed to be at the origin of the formation of the galaxies observed today. "(Cfr : Big-bang theory : Encarta© Microsoft Corporation. All rights reserved)

3.1.1.2. Our comments

For us, by analyzing these images, we can interpret the cosmological diffuse background as a trace and proof that the primordial Universe has gone through a very dense and very hot step.

In fact, according to the Universal Cosmological Model, this microwave radiation at 3 K is a residue of the energy that prevailed before the Big Bang (God says : "light be and light was"), which is an evidence of the extreme temperature that the mass of "biblical waters" had acquired just before its explosion (Big Bang) or expansion (Ge 1:2-3).

Some scholars, not least Marc Kamionkowski, Adrienne Erickcek and Sean Carroll, had a different thought than their scientific colleagues, confirming our comment about the interpretation of this image. Let's read their commentary on the second image :

"*WMap observations indicate that, on average, one half of the celestial sphere has greater thermal fluctuations than the other hemisphere... This could be a window into what happened not only during inflation but even... before the Big-bang*». (Cfr :

Futura-sciences.fr, The bizarre asymmetry of fossil radiation : an enigma solved ? Le02/01/2009 by Laurent Sacco).

Before them, other scientists had also address the same reasoning. This is the case of this article in futurascience :

"Around the 1960s, at the time of the discovery of fossil radiation, Robert Dicke and James Peebles, among other pioneers of cosmology at the time, envisaged in particular that the initial hydrogen and helium atoms came, perhaps, from an earlier step of contraction of the observable Universe (which would therefore be cyclic), and which had dissociated nuclei and electrons at high temperatures, before a new step of expansion, the one we are observing, began".(Cf : Futura-sciences.fr, First stars in the Universe : signals finally detected! On 02/03/2018, Laurent Sacco).

As the reader can see, according to the standard model, before the Big-bang, space and time didn't exist, then "boom!", a hot mass explodes as the beginning of the Universe. And, 380,000 years later, light appeared known as fossil radiation (photo and comments above).

On the other side the biblical creation scheme, scientifically restored in the Universal Cosmological Model, there was space and time before the Big-bang, even if it is not defined at the moment. This first light closes the first day of creation before the Big-bang, even if it first dissipated after the first moments of the Big-bang. So, until then, we are not contradicting science in any way.

Besides, if scientists have discovered it; it is because God through Moses revealed its eternal character by calling it "Day" (Gen 1:5) to demonstrate that it will henceforth remain attached as a seal to the Universe regardless of fluctuations. Hence its reappearance 380,000 years later after the Big Bang to this day.

So George Gamow didn't predict it, he simply recalled it, even though he was ignorant of the Bible. It's the same today where my thoughts in the bedroom in an African country leads me to the discoveries of those who use mathematics and devices. I cannot claim these personal discoveries as my own and I am obliged to use their quotations, as they offer more credit.

Creating a Universe of its own in one's room is part of the

daily life of all astrophysicists as we can read in this quote : "To found his theory of relativity, Einstein proposed to move (only by thinking, of course) by accompanying a luminous photon. » *(Cf ::http://www.matierevolution.fr/spip.php ? article38Le vide, ... pas si vide, le23/06/2010, par Robert Paris).*

To date, all the experiments carried out in the major laboratories and observations of the Universe have only confirmed his theory. The problem for us in underdeveloped countries is that when we discover something, even in these thoughts, we don't know how to spread or publish it.

When you go to a university journal, instead of being interested in your article or discovery, it asks you for your academic titles and rejects you if you are not an assistant, a supervisor or a professor. This creates the blockage of independent researchers, and therefore the blockage of solutions to some of the problems of mankind.

If it turns out that the cosmological diffuse background is the light before the Big-bang to which it is said : "God said that light be, and there was light ", this will mean that certain particles, even elements called "waters" by Moses, were indeed formed before the Big-bang, as Robert Dicke and James Peebles affirmed when they discovered the cosmological diffuse background, namely :

"The initial hydrogen and helium atoms may have come from an earlier step of contraction of the observable Universe (which would therefore be cyclic), which had dissociated nuclei and electrons at high temperatures, before a new step of expansion, the one we are observing, began". (Cfr : Futura-sciences.fr, First stars in the Universe : signals finally detected! On 02/03/2018, Laurent Sacco).

Therefore, the study of these fine rays will reveal many of the codes of the early Universe on the Pre-Big-bang period and this will mark a great step forward for Moses' field as the search for the Pre-Big-bang evidence.

As we have already said it, all radiation carries the message of its emitting source. Now, in the verses "*the divine energy (a wind from God) moving over material (waters) and light appeared*" (Gen 1:2-3), should provide information not only about that material, but also about the energies that were involved.

Today, according to several scientific sources, the glow from the Big Bang has been immensely diluted by the expansion of the Universe. For scientists, if the contribution of distant stars and galaxies is removed, the background of the heaven will become extremely dark, but there will still be some residual radiation, invisible to the naked eye : this is fossil radiation. This is one of the many reasons why we believe that this radiation is the light which the Bible spoke about. But as I said, only science can confirm it.

How do the scientists themselves explain the dark period that followed the Big-bang until the reappearance for us, 380,000 years later, of the first light that will become fossil radiation or the cosmological diffuse background? The elements of the answer can be found in the scientific documentary "*And the light was : or the story of the Big-bang*", a conference held by French nuclear physicist and astronomer Jean-Pierre MARTIN (www.Planetastronomy.com).

– An incredible density of material was found at one point (Ge 1:2);

– The temperature is enormous (Ge 1:5);

– The energy density is so great that the material trying to create itself is immediately destroyed by the radiation;

– There is no light and photons cannot escape because the Universe has become so opaque.

And as a conclusion, the speaker said : "this is only a position confirmed by the facts today, but which can be questioned tomorrow".

In the above comments, prevented by the "opacity of the Universe", does not mean that there was not yet light.

3.1.1.3. The cosmological background; a remnant of the pre-big bang

The cosmological background, also called the fossil radiation, is the first light that appeared in the universe. Why the fluctuations of this light filling all the universe became the witnesses of its origin for the supporters of the universe of before Big-bang and those of after Big-bang ? Is there a similarity between this light which appeared 380.000 years after the Big-bang and the light which is evoked in the Bible in the verse "God says that the light is and the light was ? (Gen 1:3).

We said that all the phenomena reported in the first biblical day took place in the pre-Big Bang period. Scientific studies, on the other hand, have shown that the first light called the cosmological background appeared in the post-Big Bang period. It should be noted that it is the properties of this light that came to confirm the Big-bang. The remnant of this light fills the universe to this day with a very homogeneous electromagnetic radiation observed in all directions of the heaven.

This first light does not come from a star, but from a great heat whose temperature decreases as the universe expands. What event produced this great heat after the Big-bang except the Big-bang itself? None. This is to say the explanation that scientists propose concerning the origin of this light is a trial and error because this light did not appear 380.000 years after the Big-bang, but, we say that it had rather reappeared. This is to say that its origin goes back to before the Big-bang and would be effectively the "cry of God" according to the expression of an astrophysicist who did not know that he had predicted something true.

Indeed, the first cry of God mentioned in the Bible is indeed "God said ; let there be light and there was light". But, did this light really appear by a magic cry of God ? No. The expression "God said" means that it was part of the scheme of the universe whose author, the scholar Moses, attributed its creation to God. Thus, all the breaks of symmetries had for expression "God says", whereas science speaks about "chance or force of nature".

Moreover, even astrophysicists do not often speak about the origin of the cosmic microwave background, but rather about its appearance 380,000 years after the Big Bang. This light was the physical consequence of the biblical verse that says 'The earth was formless and empty : there was darkness on the surface of the deep, and the spirit of God moved over the waters' (Gen 1:2). The scientific explanation of this verse is found in the theory of the strings of the scientist Gabriele Veneziano, which announces itself as follows : "The universe began in an infinite void (the earth was formless and empty, according to the Bible). And, in the course of its evolution, the forces of nature (a spirit of God according to the Bible) have merged the particles (waters according to the Bible), until ending in a gravitational collapse becoming a black hole (an abyss according to the Bible)".

All scientists know that the contraction of gases in space produces heat and eventually emits light. This is what the scientist Gabriele Veneziano has omitted to mention in his theory, while Moses has mentioned it in the Bible by the verse "God said that the light is and the light was", demonstrating the great heating during the processes of contraction of the particles that made the light emerge. These processes of contraction and heating continued until even light was confined in the condensed universe as in a black hole. It will only reappear 380,000 years later when the universe has expanded again.

This is the origin of the cosmological background or the first light that the American scientists Penzias and Wilson discovered in 1964.

3.1.1.4. Comments close to the Biblical Model

The comments below are close to the biblical model which closes the era minus one of the Universe by the appearance of light followed directly by the Big-bang or the separation of the waters from the waters. Let us read :

Ge 1 : 3 God said : Let there be light ! And there was light... That was the first day.6 And God said, Let there be a firmament between the waters, and let it separate the waters from the waters.

In the scientific documentary "The Universe – The Mysteries of the Big Bang", written by Laura Verkland Armstrong, we noted that "380,000 years after the Big Bang, the cloud has risen to allow us to see the structures. But we believe that these structures were created in a fraction of a second after the Big Bang.

Other scientific comments that support it :

"This signal is said to have been emitted 380,000 years after the Big Bang, when the temperature of the radiation was too low to ionise the hydrogen atom, when the radiation was said to have

decoupled from material. Doroshkevich and Novikov assumed that the radiation was in thermodynamic equilibrium with material before decoupling, and the expansion stretched the spectrum. They concluded that fossil radiation must have had a black-body spectrum. (Cfr : Techno-Science.net. By Bongo, 23/03/2014).

Science in the Standard Model has learnt to speak only of the post-Big-bang period as the beginning of everything. This is why in the scientific documentary "The Universe – The Mysteries of the Big-bang", written by Laura Verkland Armstrong, the speaker mentioned "the fraction of a second after the Big-bang". All scientists know that those who talk about the before Big-bang period attract the lightning bolt of the scientific world, because it cannot be scientifically proven. So if scientists didn't know where to draw the line on the appearance of fossil radiation, Moses did.

Moreover, the scientists themselves rejected the biblical phrase "God says" because, they say, in this sentence, the notions of space and time do not exist. And for the Pre-Big-bang period or the Big-bang period, saying that space and time were not separated, they have just unknowingly fallen back into this sentence of "God said" confirming the verse describing the Big-bang : *"God said, Let there be a firmament (expansion) between the waters, and let it separate the waters from the waters"* (Gen 1:6).

However, for us, as soon as a thing is formed, time already begins to count and the mass of the Big-bang did not appear spontaneously, reaching an opaque density at the same time and "Boom» !

Another fact is that when science tells us that the first light in the Universe appeared 380,000 years after the Big Bang, it confuses the layman in astronomy. Indeed, as I said it before, at that time, the notion of time, as we count it today did not exist because there was no reference star, from which Moses spoke of a day for the passage of one event in relation to another and we agree with him. For this evolutionary period after Big-bang, through the formation of the first generation of stars, until the achievement of what God expected, that is to say the formation of galaxies according to the Anthropic Principle, Moses had to contain it between an "evening and a morning".

Finally, in my research, I came across the words of a French scientist who confirms 100% what I say about the appearance of fossil radiation before the Big Bang. It is the astronomer Jean Heidmann. These words were taken from the programme produced on 06/06/1979 by TF1 (Collection : In pursuit of the stars). Let's listen to his answer about fossil radiation :

"And this radiation, which is currently observed in the form of radio waves, a few centimeters in wavelength, can be interpreted quite naturally in the framework of the Big-bang theory as being the relic, the relic of the very high temperature radiation that existed when the Universe was very condensed and very hot. So we can imagine that at the beginning, some fifteen billion years ago, the Universe was very hot because it was very condensed and therefore it was filled with even stronger radiation... And then afterwards the Universe remained lethargic in the form of diluted gases in completely cold space where nothing would have happened at all for the rest of eternity if there hadn't been the law of gravitation. At that moment, by chance, there were fluctuations in density in certain places, they concentrated and as they concentrated they heated up and as they warmed up they reached a temperature sufficient to ignite thermonuclear reactions". And, the light reappeared according to our conclusion".

For me, these remarks, discovered on 30/01/2018, closed and opened the debate if the radiation is a light from before or after Big-bang. The astronomer Jean Heidmann, whose words I don't know if I'm interpreting them wrongly, shows us that it is a Pre-Big-bang radiation confirming the biblical sentence "*God said that Let there be light, and there was light* ", before the "separation of the waters from the waters" or Big-bang.

Thus, the pattern described by Moses in the Bible makes him the scholar of scholars because the best that people can accept, even non-believers as scientifically rendered in this book, a pattern called the "Universal Cosmological Model". The biblical scheme of creation thus becomes, and already was, the scheme by excellence in cosmology. All believers should applaud this because our Heavenly Father, God, has just regained his place as Creator and owner of the Universe in people's hearts.

Section 4
Fossil radiation is biblical

4.1. Detecting the light before the Big Bang period

"...God says : Let there be light! And there was light" (Gen 1:3).

As I said, the fossil radiation is that light or energy that appeared before the Big Bang according to the biblical verse *"...God said, 'Let there be light'! And there was light "* (Gen 1:3). All his research or discoveries before the Big Bang we have called it "the search for the field of Moses".

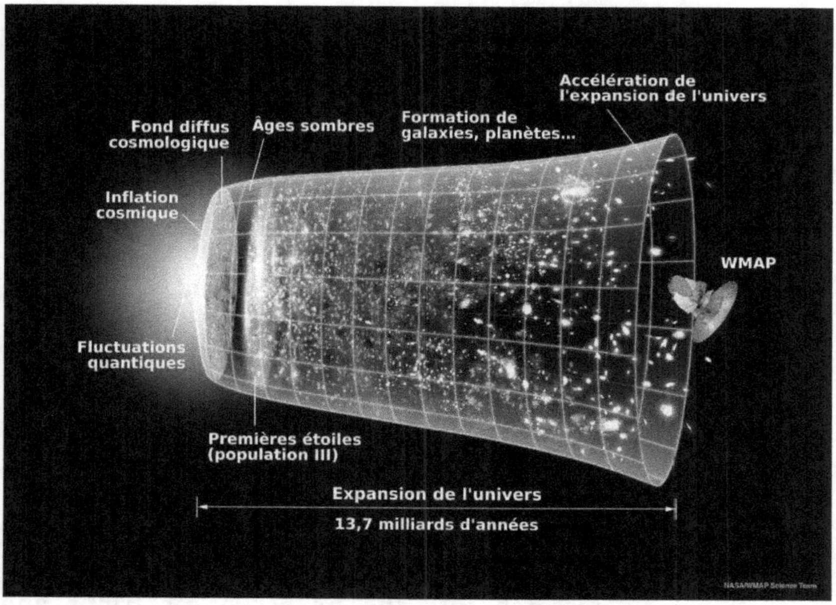

(Futurasciences.com/buildsv6/images/mediumoriginal/4/3/434bad3ced_65472_intro.jpg © Public domain).

According to the Universal Cosmological Model, the first light of the Universe was emitted before the Big-bang or the separation of water from waters. Let's call this X-energy before science confirms our statement.

So, in Moses' Field we will first look for the ray or X-energy that holds the key to explaining many of the physical riddles of the Pre-Big-bang Universe. For us, we confirm that this exists and is detectable. This X element has already been predicted by Moses and described as follows :

Ge 1:2 ...The spirit (a wind) of God moving over the waters... 3 God said, "Let there be light"! And there was light. 4 And God saw that the light was good; and God separated the light from the darkness. 5 And God called the light day, and he called the darkness night. And there was evening and there was morning : and it was the first day".

The word "spirit" is written in the Louis Segond version and "wind" is in the Jerusalem Bible. For scientists, it is the word "wind" that is most appropriate, while for theologians it is the word "spirit", but the real scientific term was first revealed by Rev. Dr. Sun Myung Moon as the Universal Prime Force emanating from the Universal Prime Energy.

We have already given explanations on several occasions concerning this enigmatic verse of an apparition of light without stars and where studies should be made. But, here we want to show why the detection of this light that appeared before Big-bang is very important for cosmology and will correct several theories. In addition, the detection of this light will answer several questions that still leave to some scientists those skeptical questions are :

- What is the force that provoked the Big-bang?
- When did dark material and dark energy appear?
- And what is the nature of the dark energy that separated this X-Energy?

If studies or research into the impact of neutrinos and gravitational waves on the cosmologically diffuse background Universe cannot answer these questions, then science, which is lagging behind religion, still has a lot to do. It has to pierce the original bubble to discover this new X-ray that was the first light of the Universe. From now on, the starting point for cosmological studies becomes the Universal Cosmological Model.

In addition, in the scientific documentary "What do you think of Stephen Hawking's statement", astrophysicist Hubert Reeves said "*science can tell us how it works in the Universe, but it cannot tell us the interpretation of the facts we observe*". However, when the light or energy of before the Big Bang appeared, a very important action was described that science is called upon to interpret *"God separated the light from the darkness"* (Gen 1:4).

Therefore, it is the discovery of this light that will also make it possible to understand darkness or everything that astrophysics calls black such as dark energy and dark material.

The energy of the first light before Big-bang also fills the Universe like dark energy and is active. In the Bible, God called this pre-Big-bang light "Day" for two reasons : to demonstrate its eternal character and its visibility (even if it takes devices to detect it) in relation to dark energy, which was called darkness.

Therefore, the word "black" that science uses to say that it is invisible was already used by Moses through the word "darkness" and Empedocles described it since antiquity as "Discord", to refer to the separating energy.

How can we detect these Pre-Big-bang energies ? Based on the Principle of Creation, a chapter of the Divine Principle of the Rev. Dr. Sun Myung Moon, which states :

"For a being to exist, energy is needed, and this energy can only be produced by the action of giving and receiving. However, no reciprocity is possible without those two elements. A movement in a straight line (the rays for example, as I say) cannot be maintained perpetually. In order for a being to have an eternal nature (including energies), it must describe a circular movement; for a circular movement to occur, an action of giving and receiving is necessary». (Cfr : Presentation of the Divine Principle, ed. Culture et paix, Paris, Pp 42-43).

Therefore, suspecting the existence of a dark energy that fills the Universe and which scientists are unable to detect, on the basis of the above law, astrophysicists studying the question will do better to track down the energy that sustains the action of giving and receiving through this dark energy and gives it the strength to act; this is the "light" before Big-bang. It is like the existence of tornadoes and hurricanes, which can be explained by the ascending and descending movement of hot and cold air.

Throughout this book I have said that the best way to detect dark energy is to first discover the ray that appeared before the Big-bang, which is this biblical "light". The problem is that we only track what we suspect, while there are things that escape us. But, once put into the paintings of suppositions, man always manages to complete his research and to lift the veils on the shadowy areas of the Universe.

Today, this biblical light has just been put into the tables of scientific suppositions called predictions, although for believers, this is not a prediction, but a reality. We do not condemn scientists, rather we leave it to them to confirm the scientific facts described in the Bible instead of imposing faith on them.

Finally, a final confirmation to close this file, specifying that the cosmological diffuse background appeared before the Big-bang, we draw it from the documentary "Cosmos – Beyond the Big-bang" written by Matt Hickey. It says this after the Big-bang : "380,000 years later, light travels through the darkness and the burst of radiation discovered by Penzias occurred then".

So when did the light that travelled in the penumbra after the Big-bang occurred, when it was already confined or present in the condensed Universe ? The answer to this question, which is part of everyone's meditation and will one day, be confirmed by science, makes the Universal Cosmological Model, which is a scientific restitution of the biblical scheme of the creation of the Universe, a Universal Model. Moses as the first to describe the pre-Big-bang period, all research based on this period is now called, we have said, "Moses' Field Research".

As a conclusion, we can say that the fossil radiation was a light that had appeared in the Pre-Big-bang period which got stuck in the condensed Universe to reappear again after the Big-bang, demonstrated in a scientific way by Penrose and Minkowski through Minkowski's Space-Time Diagram, according to the Universal Cosmological Model.

Let's take a look at this Minkowski diagram on the appearance of light before the Big Bang and its reappearance after the Big Bang.

Conformal diagram of Minkowski's space-time.

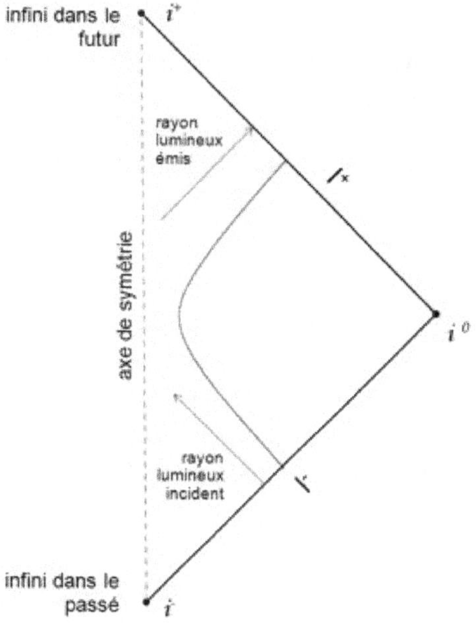

"The conforming character of this representation means that a light ray from the past follows a straight line tilted at 45 degrees from right to left and a light ray directed towards the future follows a straight line tilted at 45 degrees from left to right. The blue curve represents the trajectory of an object coming from infinity in the past and extending to infinity in the future". (Diagram by Penrose : Chaours.rv.pagesperso.orange.fr/physics/Stars/Black/penrose.htm).

Our interpretation of Minkowski's Diagram of space-time.

(Diagram of Penrose : Chaours.rv.pagesperso.orange.fr/physics /Stars/Black/penrose.htm).

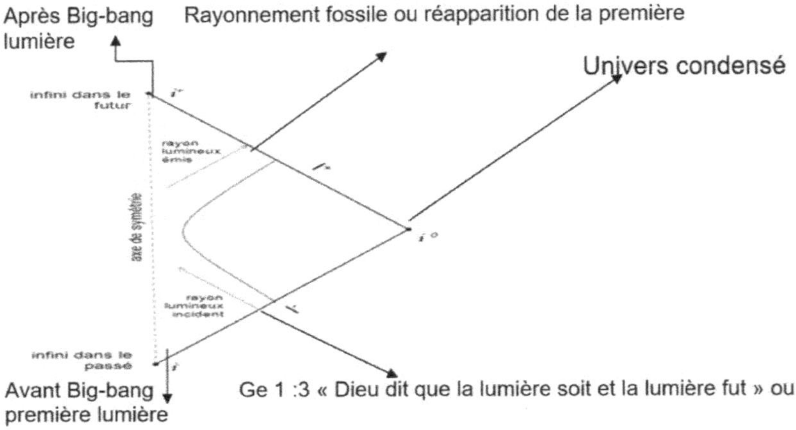

Our interpretation of the Minkowski diagram within the Universal Cosmological Model is almost the same as what their authors, Minkowski and Penrose wanted to demonstrate.

This scientific diagram shows that there was indeed an incident light ray before the Big-bang and that at the point **io** where the Universe totally condensed, nothing could be visible anymore, not even light.

Section 5
Biblical Darkness

Gen 1:2 *"The earth was without form and void (Universe) : there was darkness on the surface of the abyss".*

5.1. Olbers' paradox

Olbers asked himself the question : "If the Universe is full of stars, where does the dark night heaven come from then» ? For him, if the Universe was infinitely large and homogeneous, the heaven, whether nocturnal or not, should be as bright as the sun ; which is clearly not the case. This leads to a paradox, named after Heinrich Wilhelm Olbers, who wrote about it in the 1820s. In cosmology, it took more than a century till this assumed paradox found a real solution.

The explanation adopted at the end of the 20th century was developed by the American astronomer Edward Harrison in the 1960s. For him, according to the Big-bang theory, the Universe and the stars it contains did not always exist. And since the speed of light is constant (300 000 km/s), as a result of the expansion of the Universe, the radiation from the most distant stars has not yet reached the Earth, while some nearby stars have already stopped emitting, as the lifetime of a star is finite. With this last assertion, for astrophysicists, the Olbers paradox is resolved. (Cfr : "Olbers, paradox of."). Microsoft® Encarta® 2009 [DVD]. Microsoft Corporation, 2008).

Within the framework of the Universal Cosmological Model, this question, known as the Olbers Paradox, was already solved in the biblical scheme of creation by Moses 1,500 years BC. For modern cosmology, the before Big-bang is just an aporia, it never existed, so the American astronomer Edward Harrison took the period of the Big-bang as a reference.

But in the Bible, in the period before Big-bang, we read : "*The earth was formless and empty (Universe) : there was darkness on the surface of the deep*" (Gen 1:2). So, we see that before the Big Bang, before any creation of the elements, the state of the Universe was already dark as mentioned in the bible. If the American astronomer Edward Harrison was reading the Bible, he could have started by referring to this verse and ended it with his answer that, due to expansion, even the stars that are created and die are not able to fill or dissipate this original darkness.

As we will see in the following lines, before the Big-bang, the state of the Universe was dark. It is on the basis of the movement of a wind from God over the waters (original particles), that there was the first light matched by the verse "*God said : Let there be light". And there was light* " (Gen 1:3).

But, as before the Big Bang, the Universe was in a state of deflation, condensed to the extreme, even this light could not escape it. This is the period of confinement. Only after inflation, when the Universe began to vomit up everything it had created as particles before the Big Bang, did light also reappear in the darkness 380,000 years after the Big Bang (the separation of water from water) according to the calculation made by astrophysicists. Thus, it is inside the darkness that things are created as confirmed in the scientific documentary "The Mystery of the Universe : Light and Darkness".

Chapter III
The Big-bang

https://i.la-croix.com/729x0/smart/2014/11/14/1237059/gravure-colorisee-XIXe-siecle-apres-Raphael-1483-1520_0.jpg

Gen 1:2 "The earth was invisible and empty (Universe) : there was darkness on the surface of the deep and the spirit (a wind) of God moving over the waters. (Septuagint version)

Gen 1 : 6-8 "God said, 'Let there be an expanse between the waters, and let it separate the waters'" or (Big-bang according to the scientific name of this phenomenon).

The Big-bang, which we explained at length in Volume I, represents the zero moment when the highly condensed Universe exploded. Nowadays in science, we no longer talk about the explosion but about the expansion. In the biblical scheme of creation, this is represented by the phrase "the separation of the waters from the waters" (Gen 1:6), showing that there was something homogenous defined as the waters and which separated. The waters below serve as the creation of non-luminous spatial things, including the earth and the waters above serve as the creation of luminaries, i.e. all those who emit light.

Gen 1:2 "The earth was invisible and empty (Universe) : there was darkness on the surface of the deep and the spirit (a wind) of God moving over the waters. **(Septuagint Version)**

Ge 1:6-8 "And God said, Let there be a firmament between the waters, and let it separate the waters from the waters. 7 And God made the firmament, and separated the waters that were under the firmament from the waters that were above the firmament. And it was so.8 And God called the expanse heaven. And there was evening and there was morning : the second day.

In the biblical scheme, whose author did not have our current scientific vocabulary, we are told in the first step that "the spirit of God or a wind of God moved over the waters", which scientifically was the stage of the contraction of energies. After this was contracted to a minimum point, the second step was "the separation of the waters from the waters", that is, the Big Bang. This is not concordance or a game of vocabulary or association of words. It is indeed the steps that the universe has gone through. After that, the separation "of the waters from the waters" was followed by the expansion of the universe, which the Bible calls "expansion". This is the third step which is the one after the "separation of the waters from the waters" or after the Big Bang.

Section 1
Observational evidence of the Big-bang

1.1. Biblical and scientific evidence

People want to ask me for the observational evidence. Here it is :

Gen 1:6-*9God said, Let the waters under the heaven be gathered together in one place, and let the dry land appear. And it was so. 10God called the dry land earth, and he called the mass of waters seas. God saw that it was good.*

Gen 1:*14 God said, "Let there be lights in the expanse of the heaven to separate the day from the night; let them be signs to mark the times, the days and the years; 15 and let them be lights in the expanse of the heaven to give light on the earth. And so it was. 16God made the two great lights, the greater light to rule the day, and the lesser light to rule the night; he also made the stars.*

The Bible says that with the waters below "the Earth emerged" and with the waters above, the lights were created. And in these days scientists confirm that both the earth (formed by the waters below) and the luminaries (formed by the waters above) are composed of the same elements. The difference is only around their states : solid; liquid and gaseous for the earth and some planets, gaseous and dusty for some planets and nebulae, fluid and plasma for the Sun and other stars.

In conclusion, the first biblical day gathers the phenomena that took place before the Big-bang. In the second biblical day, we enter the step of the condensed universe and the Big-bang. And the third biblical day marks the beginning of the post-big-bang step until the formation of man. Therefore, nothing that scientists have discovered up to now contradicts the biblical creation scheme. We have already clarified the biblical days that are different from our ordinary days.

I am the first to make these discoveries after 10 years of independent and objective research. My initial motivation was only to satisfy my curiosity and to verify why the biblical scheme of creation was never taken into account in science, nor even in the history of cosmology. My work has just put an end to the rejection of this scheme.

I am also the first to convince the whole world with a simple observational proof that the Big-bang is a real fact, by the scientific discovery of the homogeneity of matter in the universe.

1.2. The Big Bang is a rebound

Ge 1 : 6-8 " *God said, "Let there be an expanse between the waters, and let it separate the waters from the waters." 7 And God made the expanse, and separated the waters that were under the expanse from the waters that were over the expanse. And it was so.8 God called the expanse heaven. So there was evening and there was morning : this was the second day.*

According to the Bible, i.e. according to the reality as rendered in the Universal Cosmological Model, the Big-bang is not the beginning of the universe, but a rebound as confirmed by the above verse. In this verse, we see that the phenomenon that science calls the Big-bang took place on the second biblical day.

This phenomenon finds its essence in the first day when it was written that "A wind or a spirit of God moved on the waters" (Gn 1.2). As we can see, the scheme of the biblical creation follows a logical path from the origin to man as rendered in the Universal Cosmological Model.

Thus, the Big-bang not being the beginning of the Universe, this secondary or transitory stage was qualified scientifically as a rebound as advocated by the German physicist Martin Bojowald. (Cfr : Martin Bojowald, l'univers en rebond, Folio Essais N° 575)

If the Bible is the only book that gives a complete scheme of the Universe as restituted in the Universal Cosmological Model, on the other hand, on the side of science, we have to put back together several pieces to bring out a scheme of the evolution of the Universe.

According to the Rebound Model, the universe, which originated from an infinite void as pointed out by the Italian physicist Gabriele Veneziano and the Bible Septuagint version, then experienced a contractive step called cosmic deflation in the Universal Cosmological Model.

While contracting according to the Universal Cosmological Model, the heat was only increasing until the light known to this day under the name of the fossil radiation. And, having reached an extreme condensation, the light could not escape. This is how it experienced the rebound, known as the big bang, followed by inflation and the reappearance of fossil radiation or light 380,000 years later.

Section 2
The energy of the Big-bang

1.1 Scientific interpretation of the activation forces of the Big Bang

In the Ekpyrotic Model, it is written as following : "The branching model may give rise to unconventional models concerning the Big-bang". In this model, the Big-bang would be a monumental exchange of energy that would occur when two branes touch. One transmits a phenomenal amount of energy and material in highly condensed form to the other and the other does the same. It is this exchange of energy that would be the basis of the Big-bang according to this theory. (Cfr : www.Wikipedia ;Brane) – (Big-bang and expansion of the Universe).

In the Universal Cosmological Model, on the other hand, referring to the biblical verses for the Pre-Big-bang, we find three acting forces instead of two as in the Ekpyrotic Model. Let's read :

Ge 1:2-4 "The earth was formless and empty; there was darkness on the surface of the deep, and the Spirit of God moved over the waters. God said, "Let there be light. And there was light. And God saw that the light was good; and God separated the light from the darkness.

In these verses we have the presence of the great energetic forces that came into action : "The Spirit of God" moving the material, the appearance of a light that comes to activate the darkness that was previously latent.

I was wondering how three different manifestations of energy could come together in the same field. I couldn't find the answer in astrophysics, but I discovered it in chemistry in Eyring's theory. It is called activation energy (Ea).

Applied in the field of physics, this discovery allowed me to confirm my conclusions about the energy that triggered the Big-bang, according to what is written in the biblical scheme of creation. For science this question still open.

In fact, to answer the question of where the energy that triggered the Big Bang came from, quantum physics teaches that there are other forms of energy that are called energies from the vacuum, according to the words of the astrophysicist Hubert Reeves in the documentary "Where does the energy of the Big Bang come from".

It is this quantum-type energy, he continued, which has been transformed into heat. Here we are indeed in front of the biblical phrase "*God says : Let there be light! And there was Light*" (Gen:1:3). This verse shows the state that the Universe had reached before the Big-bang, namely extreme heat according to the statements in this documentary.

In addition, in astrophysics, when we talk about the energies of the vacuum, we actually see what is invisible or what can be hypothetical at the moment. This is what we have highlighted in the biblical verses above, namely : Spirit of God, Light and Darkness.

As with the Big-bang, the Bible still confirms the hypotheses of the scholars. How was the transformation of quantum energy into heat (And light was) the cause of the Big-bang ? I speak of the cause because we are in the scientific field, but for us believers, everything happened according to God's will.

To answer this question, I was inspired by Eyring's theory of activation energy (Ea). But what is meant by activation energy according to science and which can enable astrophysicists themselves to formulate better theories than I do ?

Activation energy is a chemical concept introduced in 1889 by the Swedish scientist Svante August Arrhenius, which describes the evolution of a chemical speed reaction with temperature. In a way, it is an additional energy that has to be brought to a chemical system that already has energies for the reaction to take place. This energy must be of a thermal nature, i.e. related to heat.

In Eyring's theory (known as the activated complex theory), the activation energy is the free enthalpy that separates the energy of the transition state from that of the initial state. (Cfr : Wikipedia : Activation energy).

The enthalpy is defined by the expression : $H = U + PV$ where U is the internal energy of the system, P is the pressure and Vau is the volume of the system.

All this is schematized and explained in the image below taken from Encarta :

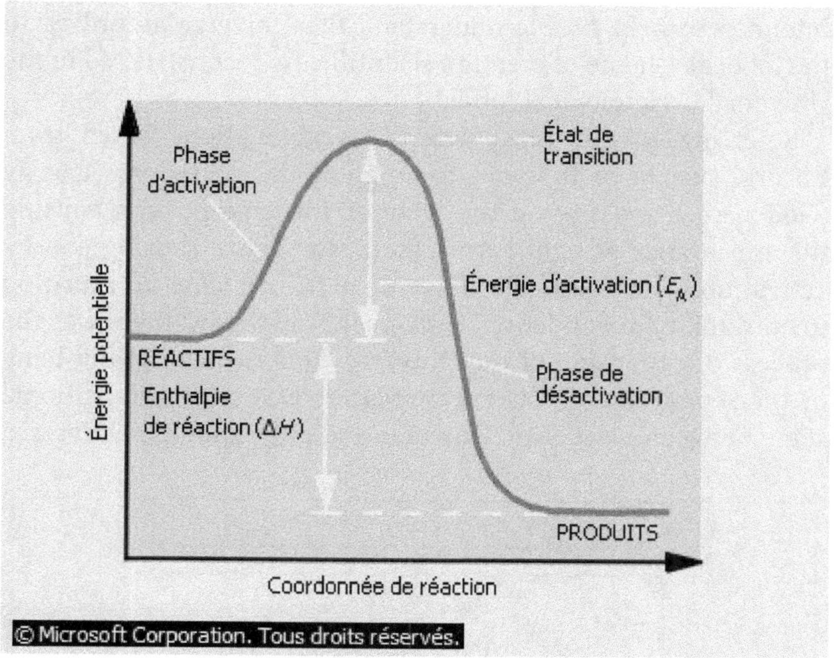

Energy profile of an exothermic elementary reaction

"This global variation in potential energy corresponds to the variation in enthalpy between reactants and products, called the enthalpy of reaction (ΔH). Furthermore, to pass from the initial state to the final state, the system must acquire a minimum amount of energy (activation step) to initiate the reaction : this is the activation energy (E a). The system thus passes through a maximum of energy and is in a state known as the transition state, which corresponds to the moment when the reagents form an aggregate, called the activated complex, in which the bonds are breaking and forming. Then the system evolves towards its final state (deactivation step), whose potential energy may be lower than the initial one (exothermic reaction with heat release) or higher (endothermic reaction with heat absorption)".

 Source : Enthalpy". Microsoft® Encarta® 2009 [DVD]. Microsoft

Corporation, 2008.

Applied in the physics of creation, astrophysicists will have a different understanding of what is explained here than chemists, and it is not only for me for me to develop it. This book is only intended to help people understand the Universe according to the biblical scheme of creation scientifically reconstructed in the Universal Cosmological Model.

Referring to the above scientific explanations taken from Encarta, we have first of all a potentially increasing energy (God's spirit moving on the waters). Extreme heating causing the appearance of light : And, there was light), causing energy activation. This causes the Universe to be in a state of transition (metastability or condensed Universe). And finally, there was the process of expansion of the Universe. This is how the Big-bang or the separation of water from waters was only a transitional stage between a deflating Universe and an expanding Universe.

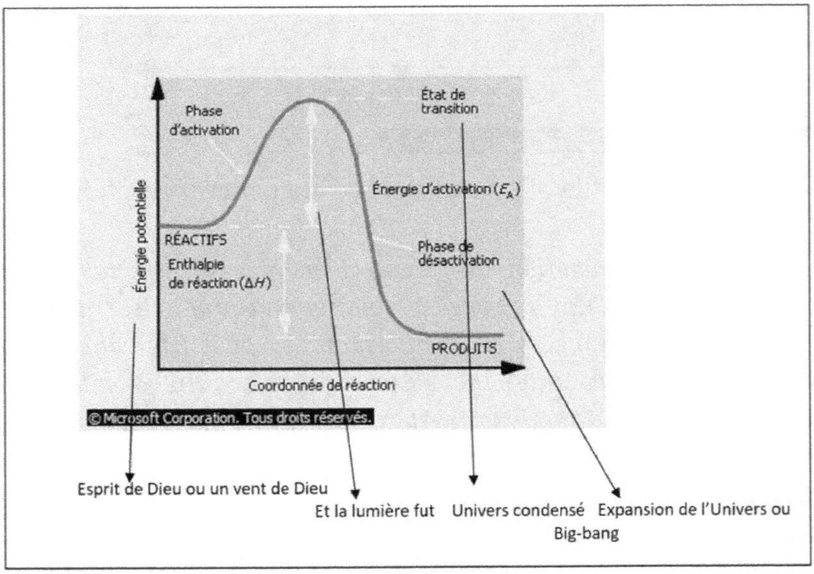

Thus, in the biblical scheme of creation, the Big-bang does not occupy the first day but the second day or the second period of the Universe according to the scientific restitution. In the diagram above, the condensed Universe occupies the summit stage, while the Big-bang occupies the beginning of the deactivation step P at the pressure and V at the volume of the system.

Let's go back to the diagram of the biblical creation scientifically restored under the name of the Universal Cosmological Model to see how things happened according to the verses below :

Ge 1:1-6 : "There was darkness on the surface of the deep, and the Spirit of God moved over the waters. God said, "Let there be light. And the light was... God separated the light from the darkness... That was the first day. And God said, Let there be an expanse between the waters, and let it divide the waters from the waters. And God made the expanse" (separating the waters).

In these successions of events, instead of God, astrophysicists speak of the spontaneous breaking of symmetry. This encourages us, however, because they recognize the existence of events in physics, having required a discontinuity causing unexplained jumps. Even biology is also characterized by many of the missing links between lower and higher species, described as a mutational accident.

Returning to the Universe, in the beginning we had to deal with passive energy "darkness" in the vacuum and active energy defined as "spirit or a wind of God moving on the waters", that is to refer to the original material. This movement expressed an energetic power that caused the cosmic deflation of the Universe, causing the collisions of the original material and its heating.

This deflation caused the temperature of the Pre-Big-bang Universe to rise. The exponential rise in temperature caused the defined original material of the biblical waters to fuse and heat up until the emergence of the first light sanctioned by the biblical verse "And God said, let there be light, and there was light".

In everyday life, when light appears, darkness disappears, but here at the very beginning of the Universe, we were only dealing with energies. So, instead of the darkness fading away, it was rather separated from the light, showing that it was not just darkness. It was this light that came to play the role of the enthalpy variation and made even the darkness become an active energy by reversing the processes after a stage of metastability of the Universe.

So through the biblical verse "God separated the light from the darkness" we see that a third energy (light) came into play by activating the darkness which came to reverse the processes after a stage of metastability and causing the Universe to enter into inflation through the process known as Big-bang by science and separation of waters by waters in the Bible.

As I said, in astrophysics, these phenomena, caused by a sudden rise in temperature creating a discontinuity, are known as the breaking of symmetry (God says), so according to quantum physics, there was already high energy activity long before the Big-bang. Without energy, the Big-bang itself could not take place. This is how some scientists say that our Universe is the result of a collapsed star in another Universe. Other astrophysicists who don't believe in the Big-bang before the Big-bang sometimes present us with diagrams where everything originated from the Big-bang.

In conclusion, the emergence of the first biblical light effectively constituted the activation energy of the Big Bang. This effectively answers the question that scholars ask themselves as to where the energy that activated the Big-bang came from.

Section 3
The stars, reflections of the Big Bang

Reflection is defined in the Encarta dictionary as a product or an indirect illustration of a phenomenon.

(Cfr : Microsoft® Encarta® 2009, all rights reserved).

3.1. Deflation and gravitation

Cosmic deflation, which is the opposite of cosmic inflation, is the phenomenon that characterized the Pre-Big Bang Universe. It is the phenomenon that allowed the original elements to merge into the condensed Universe. Stars, on the other hand, begin their formation by the fusion of the elementary elements.

We said that stars are the fruits of the Big-bang, which means that their creation follows the same process that led to the Big-bang and at the end of their cycle they emit a phenomenal light called supernova. The only difference is that the big-bang rejected the particles that had reached maturity into the Universe. This caused the elements to reconstitute themselves in too short a period of time forming stars compared to the process that led to the Big-bang.

It should be pointed out that the original material had to pass through several generations before reaching maturity, emitting its first rays, bringing to a close the minus one era of the Universe defined by the famous sentence of Moses "*God said : let there be light! And there was light* ". This light is indeed a physical phenomenon and it is up to the scientists to study it.

For astrophysicists, the sun is part of the second or even third generation of a star. If this is true, why is it not mentioned in the Bible? Because these were secondary or transitory stages in what God was aiming at and Moses only spoke of the goals that were achieved and palpable. So even science which referring to the sun as a star of such and such a generation, is astrophysics. People don't need such details. However, we are finally talking about it so that science knows that we are not ignorant and we have just brought out our justification.

One day astrophysicists will provide scientific proof of the theoretical arguments that are transcribed in the Bible, clarified in the Universal Cosmological Model. It was the explanations of the great terms contained in this work that they lacked, such as day, night, darkness, abyss, light, spirit of God, etc., that were missing. And speaking of the appearance of marine animals as the first creature, this also includes plankton, which means, archaea, bacteria, cyanobacteria, algae and going back up the chain to mankind. (Ge 1 : 20).

What scientist can still contradict the biblical scheme of creation in these days ?

To conclude on this point, until now, no element has been discovered on a planet or star that is not found on earth, casting doubt on the Big-bang theory (separation of water from waters). Thus, in cosmology, we speak of the two great properties :

– isotropy : The Universe has the same properties in all directions.

– homogeneity : material is uniformly distributed throughout the Universe.

This is what attests the Big-bang of original material for science that Moses called waters in Genesis.

What if one day we find something different from what we can find on earth, which is the microcosm of the Universe ? This would only be due to an accelerated process of mutations or evolution of the elements, due to atmospheric, climatic, bombardment or radiation conditions prevailing in the environment. Moreover, until these days, what has been discovered does not represent the totality of the composition of the Universe.

3.2. Supernova and Big-bang

The Big-bang reaction, even if scientists stick to expansion, is close to what happens to the formation and end of the cycle of stars characterized by an explosion called a supernova. This is why some scientists say that our Universe originated from a collapsed star, especially in the strings theory.

In fact, during the formation of a star according to science, the pocket of hydrogen, which is the component element of the star, reaches a high degree of compression and starts to heat up. The pressure rises until the gas gushes out. Gravitation maintains this pressure and the temperature rises higher and higher until the elements begin to fuse. Thus a star was born, becoming a natural laboratory for nuclear fusion. By releasing the energy contained in the atoms, high temperatures are produced and helium is created. This reaction releases strong energies by illuminating the surroundings. Hence the biblical verse : "Let there be light" to show that this was the goal to be achieved by God the Creator.

Still according to science, by the rotation of the elements within the star, the gravitational power compresses them while fusion seeks to make it explode. After thousands of years, i.e. after other processes that astronomers know better, which is not what this book is about, the star ends up exploding as a supernova, releasing into space the materials it has created, such as gold, silver, palatine, dust where we find the molecules of life and other elements.

So, for scientists, the man who has almost all the chemical elements is the product of the star. This was already in line with what Moses the scientist of scientists said :

Ge 1:7 "The Lord God formed man out of the dust of the earth (of the universe)".

After all, man was formed on the earth and not in a star, even though the elements were formed in the star.

Section 4
The material of the Big-bang

4.1. Plasma

What was it inside the condensed Universe? This question still an enigma for science. However, some scientists speak of concentrated energy and others suspect something resembling plasmas. We approach this question by taking inspiration from the word 'waters' that Moses in the Bible had to use.

Returning to our creation story, we set aside the assumptions and focus only on where the Bible and Science agree, i.e. there was indeed a separation, explosion or expansion called the Big-bang by Science. Nowadays, the term Big-bang sums up all this phrase or phenomena and we have adopted it in this book not as a coincidence due to chance as scientists call it, but as a result willed by God. I stress the word "result" because the process was scientific despite the phrase "God said", to show that it was part of His plan.

These events took place in the second days of creation according to the Bible, which science describes as the end of the first period of the Universe. The original material described as water in the Bible had become compact as a result of the physico-chemical reactions created by the Universal primal force. These eventually formed bubbles of energy. These bubbles of particularly intense energy were kept inside what Moses called the abyss under the action of a magnetic field until they reached an extreme density. As the heat became intense, the energy began to release itself from its constraints. The flame became brighter and brighter, transforming the original particles into plasma, and eventually emitting enormous waves of radiation (Gen 1:3 "God said, 'Let there be light'"). Sometimes held by the outer layer, confining it to an extreme density, it had an explosion or expansion, described as a Big-bang by science and as "the separation of the waters from the waters" by the Bible.

By saying that this material was in a plasma state before it exploded, contradicts scientific conclusions. As far as science is concerned, the plasma era began one second after the Big Bang. This may be true because of the confinement no light escapes. But the plasma was already in this condensation, because when it exploded. The elements dispersed and the condensation on the scale of the Universe produced a new plasma until stars are formed which can only take thousands of years. This is why some scientists do not hesitate to say that the Universe is too young to generate some of the stars they have observed. This is to say that the process was already well advanced before the Big-bang and that they only dumped the material that was already ready to generate the first atoms.

This philosophy, we have said, goes hand in hand with the cosmological thought of the 1960s as we can still read :

"At the time of the discovery of this radiation, Robert Dicke and James Peebles, among other pioneers of cosmology at the time, envisaged that the initial hydrogen and helium atoms came, perhaps, from an earlier step of contraction of the observable Universe (which would therefore be cyclic), and which had dissociated nuclei and electrons at high temperatures, before a new step of expansion, the one we are observing, has begun". (Cfr :futurasciences.com ; Laurent Sacco ; the 02/03/2018).

So, for the Universal Cosmological Model", the bubble that exploded was already plasma because, it is by reaching this level, with a phenomenal increase of temperature and not being able to support the confinement, that occurred through expansion called Big-bang.

In conclusion, for Moses, the separation took place in a material that he described as water and described it as "separation of water from waters".

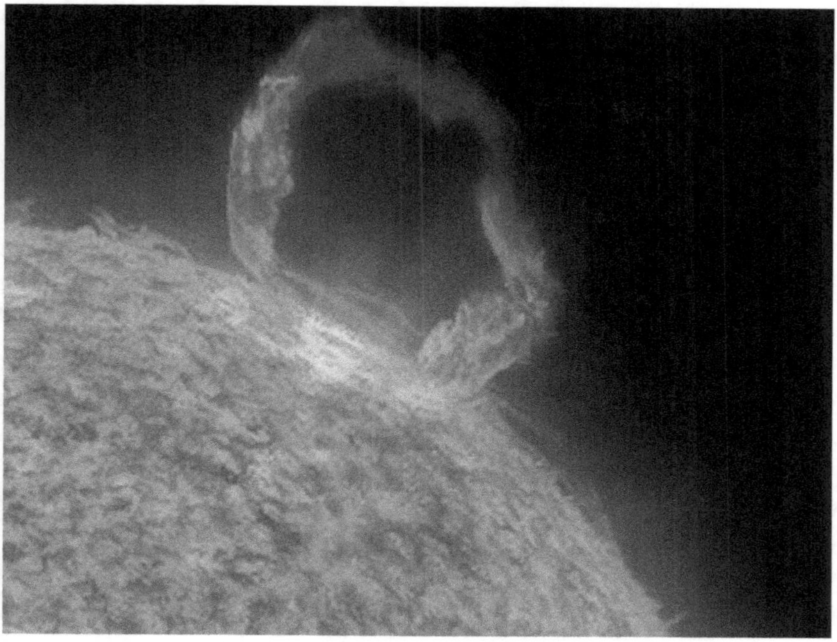

The sun is in a plasma state (Wikipedia)

4.2. The steps of biblical waters

Biblical water or pure energy went through three steps from before Big-bang to after Big-bang.

– The initial step, Moses called them "waters" were composed of the original particles.

– The second step was that these biblical waters after reaching a plasma level exploded (Big-bang) and defined by the Bible as "separation" and by some scientists as expansion.

– The third step of these biblical "waters" was the formation of the stars.

First step : original particles

Ge 1:2 The earth was formless and empty, and darkness was over the face of the deep, and the Spirit of God was moving over the waters.

Second step : plasma and Big-bang

Ge 1:6 And God said, Let there be an expanse between the waters, and let it divide the waters from the waters.

Ge 1:7 And God made the expanse, and divided the water that were under the expanse from the water that were above the expanse. And it was so.

Third step : the stars

Gen 1:9 And God said, Let the water under the heaven be gathered together into one place, and let the dry land appear. And it was so.

Gen 1:14 And God said, Let there be lights in the expanse of the heaven to divide the day from the night; and let them be for signs to mark times and days and years;

Moses explains it in literary and religious terms what the scholars explain in books that contradict each other in scientific terms.

Chapter IV
After the Big-bang

The post-Big-bang period represents the period after the Big-bang to the present day. In the biblical scheme of creation, it began on the third day (Gen 1:9). We have explained this in detail in extensively in Volume I. Here, we only want to present our point of view on certain questions, where there is a contradiction between science and the Bible in the post-Big Bang period. Some of these have already been resolved in Volume I.

Ge 1 : 9God says : Let the water under heaven be gathered together in one place, and let the dry land appear. And it was so.

Ge 1:10 And God called the dry land earth, and the heap of waters he called seas... 13 And so it was, that there was evening and there was morning, the third day.

Ge 1:16 And God made the two great lights, the greater light to rule by day, and the lesser light to rule over the night—as well as the stars.

Gen 1:19 And there was evening and there was morning, the fourth day.

Section 1
Water

1.1. The origin of water on earth

In cosmology, it has been shown that dry land preceded water. When talking about dry land, we want to make the difference between oceans and dry parts.

Astrophysicists maintain that the water in the planet Earth comes from celestial bodies such as comets and asteroids. So for them the dry part was already there. This version seems to contradict the biblical verse that says that the dry appeared in the water as we can read :

Gen 1:9 God said, "Let the water under heaven be gathered together into one place, and let the dry land appear. And it was so.

So, according to current cosmology, the planet Earth was rocky (dry) and was later bombarded by icy comets and asteroids, bringing water. However, according to the above biblical writings, the dry appeared in the water, showing the contradiction with scientists.

According to the Universal Cosmological Model, we have no evidence to support our argument. But, referring to the above biblical verse, we can say that at a certain point in time even the dry was covered with water. What we do know is that in the end it is always the Bible that wins, unless we ourselves are confused by our interpretation of the word water after the Big-bang. But, let's enter the scientists' debates on this question by reading this article from defuturascience.com dated 04/11/2014 written by Xavier Demeersman; who came to make a contribution on the subject :

"At the heart of this fascinating debate, a team of young researchers from MIT (Massachusetts Institute of Technology) and the WHOI (Woods Hole Oceanographic Institution) believe they have settled the question (on the origin of water). Their study was published in the October 31 edition of the journal Science. The answer to one of the most fundamental questions is that our oceans

have always been there (did not come from comets), argues Adam Sarafian, who conducted the research. We didn't get them from a late process as previously thought".

Another reflection recently published in Futurascience also confirms what has just been said, as we can read;

"It could therefore be argued that at the very beginning of Hadean, more than 4 billion years ago, due to much more intense volcanic activity than today, the Earth's mantle had undergone a degassing phenomenon that led to the emission of large quantities of water vapor which then condensed into heavy rainfall to form the oceans of our planet". (Cfr : www.futurascience : The Earth would have kept its water despite its collision with Theia, L. Sacco, 10/04/2018).

And a third argument :

"Researchers have confirmed the relative lack of water in the Ryugu asteroid, or more precisely of hydrated minerals, using the visible light camera and the spectrometer in the infrared pocket of Hayabusa 2. This is a surprising discovery, as it is believed that the Earth's water comes from asteroids». (Cfr : gurumed.org/2019/03/21/the-Japanese probeehayabusa-2-shows-that-lastrode-ryugu-is-tonnament-dry).

These three articles prove the Bible right when it says that the water on earth did not come from the other heavenly bodies. But therefore we must reject the argument that our oceans came from comets? Not at all, since the planet Earth itself was a BICAME (Big-Come-Asteroid-Meteorite). All it takes is a variation in the distance between the Earth and the sun to find ourselves in one of these cases.

In fact, planet Earth, enjoying an ideal position in relation to its sun, has been able to keep its water. The water that condensed in the dust disc was almost proportional to the dust. As it agglomerated in blocks, the solid mass that was created, began to float, forming the continent that science has called Pangaea. According to science "the existence of Pangaea is confirmed by the remarkable global uniformity of the fossils of terrestrial animals and plants dating from the beginning of the Triassic period". (See "Pangaea" Microsoft® Encarta® 2009 [DVD], Microsoft Corporation, 2008).

And, later on, according to science, under the thrust of the convection currents of magma in the subsoil, Pangea, the single continent, slowly divided into two parts : Laurasia and Gondwana. It is the latter that subdivided into the present continents.

Moreover, in the scientific documentary "What is the origin of water on earth", the waters found in comets have a different chemical composition from those of planet earth. They are rich in deuterium (a hydrogen atom consisting of a proton, a neutron and an electron). Therefore, our waters do not come from comets and asteroids.

A Scientific proof, taken from the same documentary, where we took a very capital element which helped us to support our arguments ; it is the crystals of Zircons.

Zircon grains, an exceptionally resistant mineral, have been found in rocks younger than 4.36 billion years old in Western Australia.

While the age of the earth is estimated from various sources to be 4.54 billion years, the study of zircon crystals from this period has revealed that it was formed in the presence of liquid water. In other words, this mineral was formed in an aquatic environment and yet its age is close to the age of the appearance of dry land. This means that water preceded dry land as mentioned by Moses in the Bible (Gen 1:9-10).

Zircon was the only fossil ore that could confirm that at that time there was only rocky magma, which was not the case. This study has changed the scientific understanding of the subject, as we can read in the following lines :

"Liquid water existed on the surface of the Earth more than 4 billion years ago. It was the study of zircon crystals that led to this discovery. These crystals were formed during the Hadean Eon, a period of 500 million years, from 4.5 to 4 billion years ago. We note the signature of a heavy oxygen isotope in zircon crystals from this period, an indication of the presence of liquid water on the surface 4.3 billion years ago. The chemical analysis of a zircon crystal (grain zircon – W74/2-36) whose age was determined at 4, 4 billion years, the oldest mineral on Earth known to date, indicates that the Earth must have already been cold enough to have liquid

water or a hydrosphere, so life may be older than we thought». (Cfr : futura-sciences.com/sciences/definitions/ chimie -zircon-4668).

Here is a proof which gives reason to Moses when he said that the dry came out of the water and shown in the following verse :

Ge 1:1:9 God says : Let the water under the heaven be gathered together in one place, and let the dry land appear. And it was so. 10 And God called the dry land earth, and the heap of waters he called seas.

By saying that the waters under the heaven are gathered together in one place, we are in the presence of the protoplanetary disc (future planet) made up of gas and dust after successive periods from the Big Bang to the supernovae of the first stars according to the Standard Model.

In the presence of magnetic interactions and very high temperatures, the gases and dusts are transformed and minerals appear by condensation. In cosmology, the word condensation is much more used to describe the physical phenomenon of the passage from a gas to a solid state and by misuse of language also to a liquid (liquefaction). (Cfr : vinsvignesvignerons.com/Geology/Une-breve-histoire-de-la-planete-terre).

The bodies continue to spin around themselves, as a consequence of God's dual characteristics, and around the proto-Sun (future sun because at that time Moses did not yet speak of the composition of the sun), attracting dust and gas present in the protoplanetary disc, becoming bigger and bigger, and forming after a thousand years rocky planetesimals or pieces of planets. The colliding planetesimals in turn release energy and gradually form protoplanets. It is the phenomenon of accretion, or growth of the planets that will lead after millions of years to the creation of the planets as the planet Earth is made up of 70% liquid water and 30% dry material, as is man, confirming the biblical statement that "man is formed of the dust of the earth" (Gen 2:7).

All these scientific processes described above are contained in the following verses :

Gen 1:9 And God said, Let the water under the heaven be gathered together into one place, and let the dry land appear. And it was so.

10 And God called the dry land earth, and the heap of waters he called seas.

As can be seen, the development of the biblical scheme of creation and its scientific restitution in the Universal Cosmological Model requires not only the assistance of the Holy Spirit but also the knowledge of the Principle of Creation and cosmology.

And, always with the participation of water and the molecules of life contained in the dust, life appeared; this is what Moses described with the sentence "God says : let the waters produce animals in abundance". (Ge 1 : 20). It should be noted that dust from the earth or stars means minerals. Out of the 92 natural elements, scientists have found nearly 70 dissolved in seawater. (Cfr : Melvin and G. Berger : What makes the ocean undulate? ed. Nouveaux Horizons – ARS, Paris2012, p.19).

Because of its ideal position in relation to the sun, liquid water on earth has been able to favor the combination of life molecules up to the appearance of life in the assembly of dust and this according to the action and plan of God.

As we said in the point about the Word of God and the meteor shower, the meteorites added to the earth and activated what it lacked for life to germinate.

1.2. The origin of water in other parts of the Universe

Gen 1:*6 God said, "Let there be an expanse between the waters, and let it separate the waters from the waters." 7 And God made the expanse, and separated the waters that were under the expanse from the waters that were over the expanse. And so it was.*

It is important to know that the word "water" was used by the translators. Moses used the Greek word hudôr, hudatos which comes from the same root; the prefix hydr. The prefix hydro– found, for example, in hydrogen (from the Greek genos, to generate : which generates water). It was Lavoisier, without knowing that he was confirming the Bible, gave the name of hydrogen to this first chemical element which is spread over the whole universe. This is attested by these first biblical verses

Gen 1:2 *The earth was without form and void : there was darkness on the face of the deep, and the spirit of God moved over the waters* (hudôr).

Hudôr means, everything that can generate water as hydrogen or everything that can be generated by water as hydrocarbon, hydraulic current, etc.

From this we deduce that the scheme of creation according to the Bible is purely scientific. It is the translators who have made it a fairy tale, as the scholars call it, because of their ignorance of the scientific terms used in cosmology, which did not even exist at that time. Therefore, one cannot blame the translators either. It is rather up to the scholars to understand at what time this scheme was revealed.

The presence of water is attested to these days, not only on the planets of the Solar System but in other galaxies as well.

In fact, the water molecule, with the formula H2O, is made up of 2 chemical elements, hydrogen and oxygen. Hydrogen is the most abundant element in our Universe ; it represents about 75% by mass of the observable material in the Universe (excluding dark material). Oxygen comes in third place behind hydrogen and helium.

Through the action of cosmic rays in interstellar media, generating chemical reactions on hydrogen and oxygen, the water molecule can form directly as ice or vapor. Thus, this molecule is present throughout the Universe, essentially in these two forms. Mixed with the grains of interstellar dust, by accretion, it can contribute to the creation of planets, comets or other celestial bodies.

Just as on Earth there are desert places without water, so is it in Heaven. If the celestial body is bombarded by its sun or star, evaporation occurs until water collects somewhere as vapors, clouds or gases, and the surface of the body becomes a desert like Mercury. But the deeper one goes into this deserted celestial body, into the unheated depth, the more tracks of water, a magmatic liquid or methane, which is often a gaseous degradation of organic material and molecules, which can be found.

If the compact or gaseous body escapes the bombardment of its star or sun, either a phenomenon of glaciations is created or the planet remains gaseous because of its remoteness. Of course, our explanation is not absolute, because for this to happen, the nebula that led to the formation of the celestial body must contain sufficient water and dust. The water having allowed the dust to agglomerate according to literary terms and at its evaporation we obtain telluric planets or meteors. However, the latter can also be planetary fragments.

On the other hand, if the nebula contains little dust, then the planet remains gaseous regardless of its distance from its star, or turns into a comet if it is very far away from it. This is the case of exoplanet 51Pegasi b, which led astrophysicists to revise their theory because they did not expect to find a gas giant so close to its star.

Section 2
God said – God did – God created and the meteor rain

2.1. Meteorites

Meteorites represent fragments of rocks found in interplanetary spaces. They can be deflected from their trajectory and fall to the earth or to another planet. When falling on a planet, if the planet has a dense atmosphere, the meteorite heats up and turns into a fireball. This is what we often call "flying stars" in common language.

In the past, these fireballs had spiritual meanings according to civilizations. For example, when Jesus was born, the three Magi used the trajectory of a meteorite to locate the place where he was born. As the name "meteorite" is a recent one, the Bible speaks of a star instead.

Matthew 2:9 And behold, the star which they had seen in the east went before them until it came to rest over the place where the little child was. 10 When they saw the star, the three Magi were overjoyed.

In Black Africa, these balls of fire have always been considered as the movements of sorcerers. In other words, these fireballs, which are only meteorites, are assimilated to sorcerers in Black Africa.

2.2. God said – God made – God created

In Volume I, we made a parallelism between the expressions "God said – God made – God created" with the breaking of symmetries. We said that these expressions were at the basis of the breaking of symmetries that evolutionists observe in living beings. In Volume II, on the other hand we will talk, in the following sentences, about the physical manifestation of these expressions in the Universe.

We know that it is impossible for science to capture the calls of God or to trace his will. But each of his words related to creation has always been followed by physical manifestations, some of which we will retrace.

Gen 1:11 Then God said, Let the earth bring forth greenness.

Gen 1:20 God said, Let the waters produce abundant living creatures ...

Scientifically, it is known that biological life originated in water, before spreading on land. When God decided to create (then God said), this event is transformed on land by certain physical phenomena. Closer to us, we have the flood and the destruction of Sodomy and Gomorrah, even though these are non-scientific events because they are not historically proven. But, further away from us, we have the meteor rain, i.e. the bombardment of fireballs on the earth.

An earth creature will say that the earth is bombarded by fire coming from Heaven. According to science, it is these meteorites that brought water to the earth and other components for life to appear. In addition, these meteorites had increased the heat of the earth so that some minerals would one day die as organic elements.

In fact, all these meteorite rains, combined with the one that detached a part of the earth that would become the moon, were not due to chance. It happened so that life would germinate on this planet.

Nowadays, as soon as water is discovered somewhere, astronomers start running to find life there. Water alone is not enough; the planet must have benefited from a standard bombardment of meteorites like the one the earth has suffered. Otherwise, it is up to man to bring life to the planet according to the verse : "Fill the universe, and subdue it" (Gen 1:28).

Let us return a little closer to the sixth day of creation. Let us read :

Gen 1:24 God said, "Let the earth produce living creatures according to their kinds..." 26 Then God said, "Let us make man in our image..." 31 ... It was the sixth day,

Chronologically, how can we date this divine way? In fact, God's spiritual creation path was followed by fires coming from the heaven, which today science speaks of as the meteor rain. What is the recent history of creation that has upset even biological mutations and allowed animals to evolve from the present mammals to the formation of man? Some scientists link this story with the disappearance of the dinosaurs.

According to one of the scenarios that scientists remember most is that of the meteorite that fell in Mexico 65 million years ago. Even the other scenarios speak more of a sudden increase in the earth's heat caused by a natural phenomenon. Based on these phenomena that allowed mammals to evolve, we can say that the sixth day of biblical creation began around -66 million years ago. This is how we understand that the Word of God is accompanied by external signs that only an exegete or prophet could interpret. Even for what happened biblically during the flood. For the inhabitants, it was a natural phenomenon, but for Moses, who revealed the providential facts of the past, it was one of the manifestations of the divine way.

2.3. The appearance of living organisms

When talking about the appearance of living organisms, we are referring to planet Earth. I had already evoked a dream I had had in a seminar when I knew nothing about cosmology. In this dream, I said that, I had seen the transformation of energies into material under the effect of lightning strikes or thunders. Throughout these writings, my vision was only based on what happens in interstellar space. What was my surprise again when I came across this book which says :

"The appearance of life on earth took place in conditions very different from those we know today. The air contained little oxygen, erupting volcanoes and violent thunderstorms were commonplace. Lightning caused simple chemicals to combine in the atmosphere, which then formed more complex chemical compounds – the basic building blocks of life".(Cfr : Linda Gamlin, The Origins of Life, ed. Artis-Historia, Brussels. 1989, P. 4).

I had also already said that, as soon as the word earth was first mentioned in the Bible, all the processes that followed afterwards were physical, according to the divine plan. That is why Moses used the expressions "God said, God did". Hence the biblical verse "Then God said, let the earth produce greenery..." which in fact represented the first molecules of life where everything that would happen afterwards was written. (Gen 1:11).

The original element that remains a mystery, to which the whole scheme was inscribed, is "water" in the verse "the spirit of God moved over the waters" (Gen 1:2).

In the above quotation from the biologist Linda Gamlin, lightning is referred to as the physical manifestation of the combination, while ignoring the inner process. Based on the Principle of Creation, by Rev. Dr. Sun Myung Moon, we discover that everything takes place on the basis of the intrinsic processes related to the Universal Prime Energy endowed on the elements. (Cf : Presentation of the Divine Principle, ed. Culture and Peace, Paris 2013, Pp 22-27).

For science, which answers only on the how, the appearance of life is part of the physical phenomena even though it is a mystery. Currently, there are several theories explaining the origin of the small molecules that led to the appearance of life on Earth. For some scientists, they could have been brought from space by meteorites (Murchison meteorite), be formed in hydrothermal mounts under the oceans or by lightning in a "reducing atmosphere" according to Miller-Urey's experiment. (Cfr : www.Wikipedia, Cells).

I had said that Miller-Urey's experience enlightened the dream I had during a seminar based on Rev. Dr. Sun Myung MOON's "Thought of Unification". When the speaker touched on the origins of the Universe, it aroused my curiosity, as it is since childhood that I intend to unravel its mystery. It was only while writing this book that I discovered that what I saw in the dream-vision which actually corresponded to the Miller-Urey experience I have already mentioned and that science should continue to investigate it.

Although I don't know what the author of the experiment meant by "reductive atmosphere", in my dream lightning created elements next to an ice cream. It was only when I wrote this book that I knew that the ice next to the created elements represented the icy atmosphere in space, without which certain particles could not remain after their creation. This was found in the Large Hadron Collider (LHC).

In this laboratory, we are witnessing the disappearance of certain particles after their creation due to the lack of a permanent temperature similar to the intergalactic climate. This is my opinion.

When I dreamt, I said, I knew nothing about cosmology or about the bodies that populate the Universe. If I had written an article at the time without any knowledge of cosmology, I would have said : God forms the elements in space by lightning next to the ice. But here, even this phenomenon of lightning is accepted by scientists and better explained by the proponents of the Model of the Electric Universe that we will see in the following chapters

2.3.1. The cyclical cycle of life

The vast majority of scientist's hypothesis think that it was meteorites, comets and asteroids that brought life to Earth. According to several scientific sources, the first cells appeared at least 3.5 billion years ago under water. The cell is the constitutive and functional unit of living organisms ; it is intimately linked to the origin of life, to the origins of the evolutionary history of living things.

All we know is that the vegetable kingdom was preceded by the mineral kingdom, because it is on minerals that we stand and the Bible states this in the verse "*man was formed from the dust of the earth*" (Gen 2:7). Man here represents not only the animal kingdom, but all life. All living things, from algae to plants to man, end up becoming dust, according to what it is written : "*You are dust and to dust you shall return*" (Gen 3:19). No living being escapes this. If Moses had written it in another book other than a religious one, a scientific law would have been detected and his name would already be inscribed among the great philosophers and scholars.

Returning to dust is in fact a cycle of life. This can be observed in the food chain of the ocean where each superior being feeds on the inferior being. Darwin called it the struggle of natural selection.

It starts with phytoplankton, which exploits sunlight and minerals in the water. The zooplankton eats the phytoplankton which is itself eaten by certain fish. The fish, in turn, serve as food for other fish. When these fish die, they become food to bacteria that add minerals to the water. As they swim upstream, the currents carry the minerals to the area of sunlight where the phytoplankton consume them to grow and revive the entire food chain in the oceans. *(Cf : Melvin and G.Berger : What makes the ocean undulate ? ed. Nouveaux Horizons – ARS, Paris 2012, p. 20).*

On earth, plants absorb carbon in the form of carbon dioxide and convert it in combination with sunlight called photosynthesis into complex compounds, which can release more energy. These compounds enter the food chain of animals, which release carbon as carbon dioxide when they breathe. This closes the carbon cycle. *(Cf : C.Bonneau, Science&Vie Hors-séries n° 288, 09/2019, p. 55).*

It should be noted that this life cycle diagram also constitutes the diagram of the implantation of life in other planets as we will explain it in Volume III.

It was in the scientific documentary "the origins of the earth" written by Alice HARPER and Nigel HENBEST that I discovered at the end of my work, the first beings were the cyanobacteria, some of them are known to have mutated into plant algae and others into fish. After fish came the flying insects, followed by reptiles, dinosaurs, primates and finally man. This is in line with the above explanations.

This scientific classification goes hand in hand with what Moses wrote in the Bible (Ge 1:20) more than 1500 years B.C. However, at that time there was no fossil study. So how did Moses know this classification that we have explained in Volume I? For us believers, the answer is very simple, the inspiration came to him from God the Creator, and what's more, he was very intelligent. Apart from this answer, can scholars offer us their answer for someone who lived at that time?

Section 3
The disappearance of the dinosaurs and the sixth biblical day

3.1. The disappearance of the dinosaurs

As I said in the previous lines, the disappearance of the dinosaurs should have something to do with the sixth biblical day. It is a serious clue to physically measure what happened at the end of one biblical day to move on to the other day.

Indeed, according to the creation story, the fifth biblical day ended with the appearance of the birds.

Gen 1:20 God said, Let the waters bring forth abundantly living creatures, and let birds fly over the earth toward the expanse of heaven. 21 And God created the great fish, and every living creature that moveth, which the waters brought forth abundantly after their kind; and every winged bird after its kind. God saw that it was good. 22 God blessed them, saying, "This is good : Be fruitful, and multiply, and fill the waters of the seas, and let the birds multiply on the earth 23 Evening came and then morning : the fifth day.

For the scientists, the dinosaurs and reptiles were the masters of this period. They would have really appeared between the first underwater beings and the birds.

According to the biochemist and biologist Linda Gamlin, amphibians evolved into reptiles during the Carboniferous period, about 300 million years ago, and these reptiles diversified. Some mutated into dinosaurs. Other dinosaurs were the ancestors of birds and mammals. A fossil known as Archaeopteryx has shown how birds evolved from dinosaurs.

The era of reptiles ended 65 million years ago, at the end of the Cretaceous period, when a large number of species disappeared as suddenly as mysteriously, including the dinosaurs. It was only after the disappearance of the dinosaurs that the first real mammals, although they appeared at about the same time as the dinosaurs, had the opportunity to grow and diversify from small nocturnal animals to present-day mammals. *(Cfr : Linda Gamlin, les origines de la vie, ed. Artis -Historia, 1989, Brussels, Pp 14-18).*

Concerning the disappearance of dinosaurs, according to the version that is accepted in science today 66 million years ago, an asteroid more than 10 km in diameter hit the surface of the earth on the Yucatán peninsula in Mexico. The shock caused a planetary cataclysm that led to a massive extinction of living species, mainly dinosaurs.

We know that the Earth does not have a perfect sphere. It is slightly flattened in the direction of the polar axes : its equatorial diameter (around 12,756 km) is greater than its polar diameter (around 12,713 km), which corresponds to a relative flattening of around 1/300. Hitting Mexico, the oceans should mitigate this impact in some areas and not all the dinosaurs would have disappeared.

For us, the shape of the earth and its diameter provide such an assertion about the disappearance of dinosaurs but can only be true if the environment was involved, i.e. the climate, or if there was an atmospheric disturbance caused by the introduction of such an object.

In conclusion, if the disappearance of dinosaurs is related to this impact, then the collision must have disrupted the Earth's orbit by causing a climatic or atmospheric disturbance that these large mammals could not withstand or that they were all thrown into the air and succumbed to the accident.

When Speaking of momentarily moving orbit, we refer again to Archimedes' Law. Since space-time behaves like a fluid with an object placed on it, if the object is subjected to momentary pressure in a very violent way in the vertical direction, it may sink further or move before returning to its initial position.

Indeed, the climate depends primarily on variations in the parameters of the Earth's orbit, which influence the solar radiation received by the Earth's surface (theory of Milanković). When there is an atmospheric disturbance, it can cause unexpected radiation to pass through despite the magnetic field.

In fact, in the conclusions that have been made so far about the disappearance of dinosaurs, and which may be well-founded, we talk about the impact, but we do not take into account what accompanies it.

The extinction theory of dinosaurs

"One of the major events leading to the extinction of the dinosaurs (as well as the 75% extinction of the of the other species inhabiting the Earth at that time) could be that of the collision of a large asteroid or comet (10 to 20 km in diameter) with the surface of the Earth (evidence of the existence of this collision has been found in Mexico). The explosion of unprecedented heat that followed the collision would have had dramatic climatic consequences on a global scale, particularly because it would have projected into the atmosphere phenomenal quantities of dust that would have shielded the sun, not to mention the probable earthquakes and tidal waves. »John R. Foster/Photo Researchers, Inc (Cf : "dinosaurs." Microsoft® Encarta® 2009 [DVD].Microsoft Corporation, 2008).

Speaking of the projection of the phenomenal amounts of dust, also confirms our assertion that many creatures have been hurled abruptly into space before falling back down and losing their lives.

3.2. Life on the sixth biblical day

We said at the beginning that the story of the disappearance of the dinosaurs should have a connection between the end of the fifth day and the sixth biblical day. On the sixth biblical day, he wrote this :

Ge 1 : 24 God said, Let the earth produce living creatures according to their kinds : livestock, creatures that crawl, and the wildlife of the earth according to their kinds. And it was so. 25 And God made the beasts of the earth after their kind, and the cattle after their kind, and every creeping thing that creeps on the earth after its kind. God saw that it was good. 26 God said, "Let us make man in our image, after our likeness, and let him have dominion over the fish of the sea, and over the birds of the air, and over the cattle, and over all the earth, and over every creeping thing that creeps upon the earth. 27 God created man in his own image, in the image of God he created him ; male and female he created them. 31 God saw all that he had made, and it was very good indeed. Evening came and then morning : the sixth day.

The physical execution of this divine word (God said), should indeed have something to do with the event that scientists are looking for, that brought about the end of the dinosaurs. A witness on the spot is said to have spoken of a fire falling from the heaven, causing the annihilation and breaking of the symmetries of certain creatures on earth. Thus no part of the earth was spared.

In view of this great phenomenon, which remains a mystery, we can say that the sixth biblical day of creation began at -66 million years ago. This reference is not a prophecy but set according to what scientists locate as the event marking the end of the dinosaurs, preceded first by the descent of fire from the heaven. The biochemist and biologist Linda Gamlin has made the following observation :

"The era of reptiles ended 65 million years ago, at the end of the Cretaceous period, when a large number of species disappeared suddenly and mysteriously, including the dinosaurs...The disappearance of the dinosaurs marked the beginning of an era rich in possibilities for mammals. The disappearance of the dinosaurs marked the beginning of an era rich in possibilities for mammals. They multiplied spectacularly, and many forms appeared». (Cfr : Linda Gamlin, les origines de la vie, ed. Artis – Historia, 1989, Brussels, Pp.14-18).

Such statements confirm and reinforce the Universal Cosmological Model. We said in Volume I, that behind the sentence "God said, God made, God created", there is always an important physical phenomenon to trigger the process. The Big-bang itself took place when God said "let us separate the waters from the waters".

Chapter V
The creation, evolution and progression of the Universe

Section 1
Creation

1.1 Conception of creation in respect of science

Let's see, if the biblical conception of creation differs from those of science

Gen 1:1 In the beginning God created the heavens and the earth.

Gen 1:2 The earth was formless and empty, and darkness was over the face of the deep, and the Spirit of God moved over the waters.

Gen 1:3 And God said, Let there be light. And the light was (cosmic ray)

Ge 1:6 And God said, Let there be an expanse between the waters, and let it divide the waters from the waters (Big-bang).

Gen 1:9 And God said, Let the waters under the heaven be gathered together into one place, and let the dry land appear. And it was so.

Gen 1:10 And God called the dry land earth, and the heap of waters he called seas. And God saw that it was good.

Gen 1:14 And God said, Let there be lights in the expanse of the heaven to divide the day from the night; and let them be for signs to mark times and days and years;

Gen 1:24 God said, Let the earth produce living creatures according to their kinds : livestock, creatures that crawl, and the wildlife of the earth according to their kinds. And it was so.

Gen 2:19 And the LORD God formed all the beasts of the field and all the fowls of the air out of the earth,

Ge 2 : 7 And the LORD God formed man out of the dust of the earth, and breathed into his nostrils the breath of life, and man became a living being.

These verses, which at first did not attract the attention of religious people, are on the contrary very fantastic to know for the scholars.

In Ge 1:1, we are told about the creation of the spiritual world (the heavens) and the Universe (the earth). The verb "to create" shows that this had a beginning. The creation of the visible Universe is studied today in physics and quantum mechanics.

Hence for us they are part of the physics of creation and part of the disciplines before Big-bang.

In Genesis 1:2, above, we are told of an empty and formless Universe, composed of dark energy (darkness) and these other energies (spirit of God), actuating the original particles (waters).

As I have repeated several times, what always made my insomnia before I understood anything about astronomy was how from the invisible, visible material came into existence, despite the famous sentence "God says"; while knowing that God also works according to laws. And, in an underdeveloped country, one can only rely on book research, thinking and God himself for the answer.

What God showed me, I said, was close to Miller's experience as quoted in this book. Hence the Universal Cosmological Model incorporates the idea of the Electric Universe, not as a Model, but as a set of theories to explain the creation of the elements.

For those who want me to repeat, at the beginning of the Universe, there were electrical phenomena, which I cannot express clearly, which arose from anti-particle collisions. These phenomena provoked strong electrical discharges, the energy which, when released, transformed into radiation, particles sometimes manifesting themselves as dust and gas. The created particles, which came from the vacuum, ended up directly in the extreme cold of the Universe. Why cold? According to the scientific simulations made by scientists, without cold, many of the things created would lose traces or life if they are not directly subjected to cold.

My words can be confirmed by the following quote : *"Material is indeed produced at the interstices of vacuum bubbles. "*

(Cfr : matierevolution.fr/spip.php ?article, Les bulles de vide et la matière, 16 May 2008, by Robert Paris).

But in my dream, there were on the one hand the detonations and the creation of the elements and on the other hand the ice. This is the way God speaks. He speaks through symbols and images ; it is up to man himself to find the explanation or to devise an understandable sentence. The Old Testament is full of such examples, especially in the story from Jacob to Joseph (Ge. 37 to 41) and Daniel (Da. 2).

The image I saw looked something like this :

Keith Kent/Science Source/Photo Researchers, Inc.

The phenomenon of lightning corresponds to an electrical discharge that occurs between a cloud and the Earth, or between two clouds. (Cfr :Keith Kent/Science Source/Photo Researchers, Inc.Microsoft ® Encarta ® 2009. 1993-2008 Microsoft Corporation).

Deep-heaven phenomena like this marked the transition from the invisible to the visible by pouring dust, gas and plasma into the Universe. In this area from the dream I had, there was also a large block of ice. That is to say that after the creation of the particles, what remained was the one that was poured into the extreme cosmic cold.

According to the latest information from the media; astronomers sometimes observe lightning coming from the far reaches of the Universe without being able to explain the phenomena and the role it plays in the cosmos. Among these phenomena, we have the gamma-ray bursts.

In the above verses (Ge 2 : 7 and 19), if for Heaven and Earth, Moses spoke of creation; for animals and man, he used the sentence "formed from the (dust) of the earth". However, scholars know that the dust of the earth (Universe) begins its cycle in the nebula. For non-cosmologists, we tell them that the nebula is a concentration of gas and dust observed in the interstellar medium (between the stars), referred to in the Bible as the extended medium (Ge1:6).

To reach the formation of a star or planet, the nebula must go through many cycles over several million years. At each stage of its cycle, it is shaped, especially in stars, to transform itself to the level of life molecules. And the earth was the ideal place where these molecules actually became life given all these conditions according to Brandon Carter's Anthropic Principles and whose main idea, we have said, is confirmed in the Principles of Creation, a branch of the Divine Principle of Rev. Dr. Sun Myung Moon. That is to say that molecules of life can be found everywhere in the Universe without there being life.

There is one more very important point which earned Moses the title of Scientist of Scientists. Scientists believe that in the biblical scheme of creation there are only the verbs create, do and form without taking into account essential physical or chemical phenomena. However, Moses referred to the participation of liquid water in the activation of life. The Bible thus proves to be the cosmological book par excellence to this day. In fact, Moses spoke of water before life was born :

Gen 2:4 These are the origins of the heavens and the earth when they were created. When the LORD God made the earth and the heavens, 5 no shrub of the field was yet upon the earth, and no herb of the field yet sprouted : for the LORD God had not caused it to rain upon the earth, neither was there any man till the ground. 6 But there arose a vapour from the earth, and watered the whole face of the ground. 7 And the Lord God formed man out of the dust of the earth, and breathed into his nostrils the breath of life, and man became a living being.

In view of this, let us read the points of view of scientists on the appearance of life :

"Water ice or water vapor has no special value for life. In fact, the value of liquid water lies in its ability to promote chemical reactions. Thanks to its polarity, water can dissolve numerous molecules, which are present throughout the volume of the solution : this allows these molecules to meet within the liquid and react efficiently, which explains the importance of liquid water in the development of life" (Cfr : Liquid water, source of life in the Universe 17/01/2011 – By Vincent Boqueho).

As can be seen, this quotation effectively confirms the previous verses (Ge 2:5). As long as there was not yet water with adequate chemical proportions, life could not germinate. In the Bible, rainwater also symbolizes which had acquired chemical proportions.

Other comments :

"It would take three major necessary but perhaps not sufficient ingredients for life to appear. The first ingredient is the chemistry of carbon, which is universal since carbon is one of the elements naturally produced by the stars, so it is found almost everywhere in the Universe. The second ingredient is water, this mixture of hydrogen and oxygen is abundant in the Universe, so there is water everywhere. The last ingredient is energy, which is present mainly in the vicinity of the stars". (Cfr :10/12/2014 – By Jonathan SARE, Futura-Sciences).

The explanation of the last ingredient confirms what the Bible and the Divine Principle of Rev. Dr. Sun Moon teach. It is not the energy in the vicinity of the stars, but rather the breath of God as the Universal Prime Energy that God has endowed to everybody through the different stages of creation. Moses referred to it as God's breath. Rev. Dr. S.M. Moon was the first religious man to explain it as energy. So, the last ingredient, which I have already mentioned, is the primary Universal energy that exists in every body. The verses and the following sentence confirm this :

Ge l:2 "the spirit of God was moving over the waters";

Gen 2:7 "The Lord God formed man (mass or body) out of the dust of the earth, he breathed into his nostrils a breath of life (energy) and man became a living being".

In the Divine Principle we read : "The First Universal Force has its origin in God and is the force that God projects into every being at the moment of its creation" (Cf : Divine Principles, levels 4, P. 21).

So, when asked whether there is a difference in the models of creation according to the Bible and science, we can say that the "star dust" according to science, the "scum" of Big-bang according to Stephen Hawking, the "dust of the earth" according to the Bible, the "mud" according to the Koran, from which living beings were formed represent the same elements to which scholars have referred to as the origin and the religious at the end.

The main elements are carbon and water. Through water we have the mixture of hydrogen and oxygen which are abundant in the Universe. Thus we see that Moses did not neglect anything in the Bible when he spoke of the dust of the earth and water and that science has confirmed this.

What cosmologists have taken to be an incomplete scheme turns out to be more complete than we thought, if we put the pieces of the puzzle together very well. This is what I have managed to present in the Universal Cosmological Model.

So the biblical view of creation and the scientific view come together, even if the basic thinking differs. The dust of the earth means the minerals found in the earth. For example, the human body consists of about 10^{29} (1 with 29 zeros) quarks, the constituent particles of the atomic nucleus. (Cf. : Blandine Pluchet, la physique quantique pour les nuls, ed. First, Paris 2018, p. 154)

Section 2
The evolution

2.1. Conception of evolution according to the Universal Cosmological Model

Are living beings the product of evolution or creation ? I don't want to throw an atomic bomb at religion, nor do I want to limit science with a dry answer, by choosing one or the other assertion. The Universal Cosmological Model, as a scientific model and not faith theories, opens the way to new research. However, the biblical scheme of creation is clear on this subject when it uses the verb to create for the galaxies and to form when it comes to living beings.

Ge 2 : 4 These are the origins of the heavens and the earth when they were created. When the Lord God made earth and heaven

Ge 2:19 And the LORD God formed all the beasts of the field and all the fowls of the heaven out of the earth,

Ge2:7 And the LORD God formed man out of the dust of the earth, and breathed into his nostrils the breath of life, and man became a living creature.

In fact, in these verses Moses speaks of the creation of the earth or the creation of the elements and the formation of man out the dust of the earth (of the created elements), showing that the process of evolution was respected. From the dust at the bottom of the seas (Ge 1:10), through the cyanobacteria, the animal kingdom, until reaching the level of man, several steps have been reached and it is up to science to discover and teach us. But what was the form of this evolution? According to Wolf-Ekkehard Lönning, a researcher at the German Max Planck Institute for Phylogenetic Research : *"Mutations cannot transform an original species (plant or animal) into a completely new species.* »(Cf : Wolf-Ekkehard Lönning in "Was life created ? www.jw.org/fr).

One thing is certain ; the present form of man was predestined from the very beginning of creation, because one of the last desires of man has always been to fly like a bird. Each one of us in our childhood had this desire. But if man was to evolve in his forms by the force of thought, we should have wings today. Not having reached this capacity, we attribute it to the angels, whom we think are super beings more than we are. This leads us to ask ourselves questions about the evolution of our present form in relation to what science shows us, i.e. a man with a curved back resembling a monkey. Wasn't this form simply due to ignorance of our true physics?

For example : sitting cross-legged is an ordeal for Westerners and Africans, which is not the case for Orientals. But, for those who have practiced, they have reached the levels of the Orientals and in the long run, it becomes as natural as it is for the Orientals. Therefore, should we speak of evolution for those who have mastered it or of ignorance for those who have not? The answer is clear; it is the ignorance of our physical capacities which should be similar to cats as yogis do. This example is confirmed by the words of Jeffrey Schwartz, an evolutionary theorist, who said : *"Natural selection may help species to adapt to changing conditions of existence, but it does not create anything new"*. (Cfr : Jeffrey H. Schwartz, Sudden Origins – Fossils, Gened, and Emergences of Species, 1999, Pp. 317-320).

But, as I have said, whatever form primitive man was given, from lizard to homosapiens, he was still a man, and already endowed with the inner capacity to communicate with God and the intelligence to handle leaves, woods and stones as a tool.

This is illustrated in the following diagram :

https://www.slayerment.com/science-vs-religion-which-right #comment-9366

There is chemical affinity between all creatures because they are all made of the same molecules of life described by the Bible as "the dust of the earth". In addition, scientists minimize the differences in percentages that mark the difference between us and the apes. But these differences have not been lost, nor have they been gained by humans through a hereditary process from the apes. Therefore, we are our own pattern.

Isn't it written in the Bible that they astonished God when they recognized their nakedness at the dawn of creation and covered themselves with leaves ? Which the animals never did. Let's read :

Gen 3:7 And the eyes of man and woman were opened, and they knew that they were naked; so they sewed fig leaves together and made coverings for themselves.

Ge 3:11 And the LORD God said unto the man, who told you that you were naked ?

In fact, the man lived in the midst of animals who were also naked, where did this shame come from in his consciousness and the desire to hide his sex organs, rather than another part of his body such as the navel for example ? The Divine Principle of Rev. Dr. S.M. Moon is the only teaching that answered this question.

There is no need to look for an incomplete answer elsewhere which will create new questions.

So, I was saying that as far as evolution is concerned, whatever form science imagines or describes primitive man, it is not a problem for the religious. All we know is that man was created by God even if his behavior or form would have undergone what science calls evolution. Therefore, for the book "The Universal Cosmological Model", the theory of evolution is not related to the origin, nor the qualification of the fittest, nor intelligence and spirituality, but rather the form and behavior.

The following sentence used in several scientific books according to which "man began with an animal" should be replaced by "man began with a form animal or cyanobacteria, etc.", but, he was man created by God. To say that man began with the ape is false, because the ape is already an evolved and final stage in creation.

I also support my thought with this thought of Rev. Dr. Sun Myung Moon :

"If humans are resultant beings, is it possible that monkeys are our cause? Can monkeys be our ancestors? Let's get over with this nonsense! Starting from an amoeba, for life to have reached human form, it had to pass through the gates of love on thousands of levels. Does life progress automatically? Absolutely not. It is the same with all animals. The division of species is very strict. No one can transgress the separation of species. If materialists, convinced that monkeys are our ancestors, cross a human being with a monkey, do you believe that a new form of life will appear? This attempt is doomed to fail, even if they try to do so for thousands of years". (Rev. S.M. Moon : Speech : True Family and True Universe Centered on the Love of God. 1997).

In this same context of idea, the famous paleoanthropologist and lecturer at the College de France, Pascal Picq, in a documentary entitled "First Man", said that we do not descend from monkeys, we are monkeys. This is to say that we are the descendants of ourselves. According to this professor, we are monkeys, but the difference with real monkeys is that we have the ability to transform things, to love each other and to worship.

So, concerning the evolution of form, in the beginning, everything starts with aquatic life, if it fauna and flora, and if it's not the scientists who will contradict me, because science must do research on the statements of the religions. It is enough that all this water vapor contained in the atmosphere, although it is only a tiny percentage, becomes liquid, for us to find ourselves as we were at the beginning, as we can read in this scientific commentary :

"If all the Earth's ice caps were to melt, it is predicted that the oceans would see a spectacular rise in level of 65.8 meters...". (Cfr : trustmyscience.com/voici-a-quoiressemblerait-la-terre-si-toute-la-glace-venait-a-fondre by Julien Claudet, July 17, 2017).

But God, Our Heavenly Father, said that from now on, He will have an eye for it. So He is behind all the environmental protection laws. Let's read :

Ge 9:14 " Whenever I form clouds over the earth and the bow appears in the clouds, 15 I will remember my covenant between me and you and all the living creatures : water will never again become a flood to destroy every creature.".

As I said at the beginning, after the Big-bang, everything became resultant and we became the fruit of a resultant, that is to say, resultant beings. Therefore, no one can affirm with certainty, in spite of the remains, what the origin really was if he does not refer to the Bible. The Big-bang, or the separation of the waters from the waters according to the Bible, has marked a line on what we can easily discover in our research on the original elements of the Universe and that we will have great difficulty discovering, especially the elements before the Big-bang. Hence the conclusion of some scholars who say that nothing can be speculated beyond the Big-bang.

The Bible is the only book that enlightens us about the energies that were at work before the Big-bang. It is now up to the scholars to make positive investigations into the Biblical Models. And for this step, for the first time, scholars are called upon to humble themselves by resorting to the sacred writings by excellence for the moment, namely the Divine Principle and the Thought of the unification of the Moon couple to better understand the working forces at the beginning of the Universe.

As far as my own experience is concerned, although I have lived and grown up in the Church of Unification, not being a physicist, the explanations of the points raised in the Principles of Creation on energy and logos, materials contained in the books mentioned above, remain very theoretical, yet it is through this understanding that a true scholar should begin his research. And I never stop reading and rereading these subjects, which is what has allowed me with "one billionth of knowledge that I ask God" in my prayers to come to the conclusions contained in this book.

As for the terms sacred books, it sometimes has a pejorative character, that is to say something set apart for certain categories of followers ; the Bible for Christians, the Koran for Muslims, the Kangyur for Tibetan Buddhists.

When I speak of the sacred writings, I simply mean that the source is right, that is to say, reserved for everyone. We can talk about metaphysics. But if it corresponds to my understanding, where is the problem, if not otherwise.

The physicist Marie Curie used to say : *"In life, nothing is to be feared, everything is to be understood"*, and this is what I have always done.

2.2. Living beings are the fruits of the stars and the earth

At the time of Moses, when we look at the human flesh with all its components, it was crazy to say that it is taken from the earth and yet, the analysis of it proves that the Bible is right. By the word dust, we mean all the elements found in the universe. And in these days, science has already discovered several chemical elements, elements that are found on the earth in the composition of the human being. It is enough to see the composition of multivitamin drugs.

According to scientific sources, six elements make up 99% of the mass of the human body. This is carbon, H is hydrogen, N is nitrogen, O is oxygen, P is phosphorus and S is sulfur. They are represented by the acronym CHNOPS. The human body also contains tiny amounts of elements. These include germanium, antimony, silver, niobium, lanthanum, tellurium, bismuth, thallium, gold and even radioactive elements such as thorium, uranium and radium. (Cf : https://www.greelane.com/chemical-composition-of-the-human-bod).

Given that scientifically all these elements were first formed in the stars, this leads astrophysicists to say that man is the fruit of the stars based on the upstream. But Moses, basing himself downstream, that is to say where man was actually formed, was more precise than the scientists in saying that "man is the dust of the earth".

Gen 2:7 The Lord God formed man from the dust of the ground...

Thus to this day our organism needs all these mineral elements for its growth and lacking even one of them, leads us to an imbalance. Man needs to constantly renew these elements in his body. If it is impossible for us to consume for example iron in its raw state while our organism needs it, God provided machines which sort out these elements in the basement and transform them into organic element for the good digestion of the animals and the men according to what it is written in this verse :

Ge 1 : "*29 And God said, 'Behold, I have given you every herb bearing seed that is on the face of all the earth, and every tree that has in it the fruit of a tree and bears seed : it shall be your food. 30 And to every beast of the earth, and to every bird of the air, and to everything that moves on the earth, having the breath of life in it, I give every green herb for food. And so it was.*

It is through plants that man can consume these chemical elements for his health. This is why we see vegetarians everywhere. Therefore, plants provide us with an honorable work and we should respect them and use them only in a good way. There is nothing that grows or was created at random. The moment has come for the man to offer to the mankind what contains even the smallest grass and we will understand that the age of Adam and the other biblical patriarchs were not fables.

Section 3
The progression of the Universe

3.1 Conception of the progression according to the Universal Cosmological Model

The theory of progression concerns living beings in the Universe. This theory, according to us, is conceived in the following way : each being has its own identity. It is born as it is, but due to external environmental conditions, it can progress but still keep its identity.

This theory is in contradiction with what scientists call the theory of evolution, we have developed our point of view in the points dealing with the biblical conception of creation in relation to evolution in this Volume, and the one dealing with the progression of the human race in Volume IV.

Chapter VI
Galaxies and planets in the Solar System

Section 1
Galaxies

1.1. General

The word galaxy may seem new to my readers, as it did to me before I began writing this book. Galaxies are what we in everyday language call the world and other worlds. Our galaxy, our world, is called the 'Milky Way'; via lactica in Latin. It extends over a diameter of about 200,000 light-years. (Cfr : Wikipedia.fr, galaxies). Other worlds are other galaxies. In astronomical literature, the word "galaxy" written with a capital G refers to our own galaxy (the Milky Way), in order to distinguish it from other galaxies or other worlds.

What do galaxies contain? All galaxies contain almost the same sets and elements and the same physical phenomena as those found in the Milky Way. This confirms, moreover, that the Universe began with a condensed material that Moses and the great ancient philosophers called "waters", and which exploded; a phenomenon known biblically under the name of the "separation of the waters from the waters" and under the name of Big-bang according to science, and whose debris is at the base of every star that we see in the Universe.

All galaxies including space form the Universe. It is estimated that the diameter of the observable Universe is 100 billion light-years. It contains about 7×10^{22} stars, scattered in about 100 billion galaxies, which are organized in clusters and super clusters of galaxies. But the number of galaxies could be even larger, depending on the deep field observed with the Hubble Space Telescope. (Cfr : https://fr.wikipedia.org/wiki/Univers).

Moreover, galaxies are not only composed of stars, planets and gases. They also contain dark material, which is invisible. According to various scientific sources, our galaxy, the Milky Way, has between 100 and 400 billion stars. It is difficult to make an exact estimate, because it is constantly changing, with millions of stars being born and millions more exploding. God himself knew that man could not count them (Gen 15:5). There are about a minimum of the 100 billion planets. A planet is like our earth with its 5 continents and oceans. But there are also planets that are made up only of gases like Saturn and others of metals, stones or ice.

Our galaxy the Milky Way belongs to the local group of a cluster of about 60 galaxies measuring 10 million light years in diameter, which also contains the Andromeda galaxy. The local group is part of the Virgo super cluster, which itself includes an even larger structure a super cluster called Laniakea. Moreover, our galaxy is not immobile in space. It is moving at 630 km/s, or 630 x 60 (one minute) x 60 (one hour), which is about 2.3 million kilometers per hour. (Cfr : futurasciences.com/sciences/définitions/Universs-voie-lactée-3729 by Floriane Boyer).

Other galaxies also have a similar composition to ours. We find suns, stars, exoplanets, chemical elements and phenomena similar to those of the Milky Way. The Universe – whose true extent is unknown – could contain up to 2,000 billion galaxies, each of which could contain 1,000 billion stars. But this can only be confirmed by observations from future telescopes.

The size of the stars is not like the little balls we see in the heaven. It is because of their distance that they appear as small luminous balls. The Sun itself is a star that can hold more than 1 million times our planet Earth. However, despite its size, it is only a small star compared to the others. In our Galaxy, there is also a star that is 1,000 times larger than the Sun and it is called Betelgeuse, radiating more than 100,000 suns combined. (Cfr ; Betelgeuse : Wikipedia).

So, with such dimensions confirmed by science, philosophizing about the Universe is part of a great invention, as Aurélien Barreau said. (Cfr : Aurélien Barrau : Big-bang and beyond, cosmology walk, éd. Dunod, Paris).

Section 2
The planets of the Solar System

In our Galaxy, the Milky Way, there are planets which are under the gravitational influence of the sun, and which are called the planets of the Solar System. There are eight of them, including Mercury, Venus, Earth, Mars, Jupiter, Saturn, Uranus and Neptune. The other planets that orbit around the stars, but still in the Milky Way, are called exoplanets. I say there are about 100 billion of them. The majority of scientists believe that some of them may harbor life, or even intelligent beings, called extraterrestrials.

The majority of the planets in the Solar System are visible to the naked eye and appear as faint stars in the heaven. This brilliance comes from the fact that they reflect the sun's rays back to the earth like a mirror facing the sun while being pointed at something.

https://www.podcastscience.fm/dossiers/2019/09/30/la-naissance-des-planetes/

In the inner system there are 4 rocky planets. These 4 planets have masses equal to or less than the mass of the Earth with

5% for Mercury, 10% for Mars and 80% for Venus. Beyond that, there is an asteroid belt between Mars and Jupiter which contains about 4% of the Moon in mass, so it is not huge. In this belt swirl pebbles bigger than others, with Ceres for example, which is a dwarf planet of 1.2% the mass of the Moon.

And then come the gaseous planets with Jupiter (300 land masses), Saturn (95 land masses), Uranus (14 land masses) and Neptune (17 land masses).

Finally, there is another debris belt, the Kuiper belt, which contains between a tenth and a quarter of the mass of the Earth and also dwarfs planets like Pluto, which makes up 18% of the mass of the Moon. The asteroid and Kuiper belts are debris belts, i.e. what remains once the system is formed, without enough mass to create real planets. Pluto was excluded as a planet because a moon bigger than it was discovered.

2.1. The age of the solar system confirms the biblical scheme

The solar system contains a set of stars that are subject to the gravitation of the sun. These stars revolve around the Sun. This System, as we said, is composed, besides the Sun itself and by order of increasing distance to it, of four telluric or solid planets (Mercury, Venus, Earth and Mars), of an asteroid belt composed of small rocky bodies, of four giant planets (two gas giants that are Jupiter and Saturn, and two ice giants that are Uranus and Neptune) and of the Kuiper belt, composed itself of icy objects.

I wrote in Volume I that stars are the fruits of Big-bang, that is to say that at each explosion of a star, there is a new process of recreation of stars. This is why the ancient philosophers were able to describe the functioning of the whole universe based only on the solar system.

According to science, all the planets of the solar system were projected by the explosion of a star, from which part of the dust and gas, having formed the molecular clouds, collapsed gravitationally. 99% of the matters contained in the center formed the Sun. The remaining 1% of the matter formed the planets.

Scientists estimate the age of the Solar System to be about 4.6 billion years, which means that according to the hypothesis of its formation, none of its stars can have an age beyond this. This is where things get very interesting.

What does the biblical scheme of creation say about the origin of the solar system? For lack of scientific terms, Moses used the name "waters", and spoke of the "separation of the waters from the waters". From the waters below, the earth was formed and from the waters above, the lights were created (Gen 1:6-19). This means that the planets and the lights were all formed by the separation of the same mass of matter that Moses called waters.

Gen 1:6-9 God said, "Let the waters under the heaven be gathered together in one place, and let the dry land appear. And it was so. 10God called the dry land earth, and he called the mass of waters seas. God saw that it was good.

Gen 1:14 God said, "Let there be lights in the expanse of the heaven to separate the day from the night; let them be signs to mark the times, the days and the years; 15 and let them be lights in the expanse of the heaven to give light on the earth. And so it was. 16God made the two great lights, the greater light to rule the day, and the lesser light to rule the night; he also made the stars.

The age of the Sun is estimated at 4.57 billion years, while that of the Earth at 4.54 billion years, proving that they are twin stars. The creation of these stars very close to each other is an observational proof made by scientists themselves without knowing that they have confirmed the biblical creation scheme.

In fact, today scientists confirm that both the earth (formed by the waters below) and the luminaries (formed by the waters above) are composed of the same elements. The difference lies only in their states : solid, liquid, gaseous for the earth and some planets. Fluid and plasma for the Sun and other stars.

All, in the solar system come from molecular clouds qualified by Moses of waters. The molecular cloud is composed in great part of hydrogen molecule which means "that which generates water". So Moses in qualifying the original mass of water was not far from science, because water (H2O) is composed of hydrogen and oxygen.

Moreover, it is only to make understanding easy that the word used by Moses "waters" was translated, because the original Greek word is ὕδωρ (hudôr) which means hydro (cfr : water; wikipedia), i.e. something that generates water.

Chapter VI

The Sun or the Universe in miniature

The study and the knowledge of the Sun should lead us to know how the Universe works. Hence, we call it the Universe in miniature. Thus, in this section, we will study the sun in its entirety as the Universe in miniature. It is already known to astronomers that this is the micro-laboratory for learning about the physics of stars.

Section 1
Sun's description

(Documentary recommended : The Sun, Source of Life and Danger {by Gary Wright})

How can we understand the Universe as a laboratory? Until now, the Sun has been analyzed as a star in astrophysics and it is believed that we understand a lot, although all the bodies of the Universe still possess some mysteries. Here it is now a material of analyzing it as the Universe in miniature. And, it is through the Sun that we can understand other manifestations that take place in space.

1.1. The Sun ; micro-star and atomic macrocosm

The Belgian physicist Father Georges Lemaitre (1894-1966) said : "*a truly complete cosmogony should explain atoms like the Suns*». (Cfr : Blandine Pluchet, quantum physics for dummies, ed. First, Paris 2018, p. 224)

Indeed, if the sun is a micro star, it is rather an atomic macrocosm. It is from the sun that we can easily discover its forces, the visible and invisible particles from the atom to the Universe. Invisible elements are released after fusion and fission if possible, and can become detectable on sophisticated devices after millions of accumulation's years.

In the Bible in Romans 1:20 it is written : *"For the invisible perfections of God, his eternal power and divinity, are seen as to the naked eye, from the creation of the world, when we consider them in his works. They are therefore inexcusable".*

In Rev. Dr. Sun Myung Moon's Principle of Creation, creations in general and human in particular were taken as a case study to explain God and what is invisible in Him. Therefore, in order to understand the Cosmos, the study must be done on the integral man. The latter, as the cosmos in miniature, not only possesses the chemical elements, but also the invisible energies and material which become organic, in other words it is visible inside the cell. It is up to science to deepen the research further. He is also a spiritual being.

But as far as the Universe is concerned, how can we understand and analyze the physical phenomena taking place inside and certain phenomena which are in our atmosphere, but whose manifestation are still unknown to us? It is through the study of the only star that is close to us, namely our sun. The option has already been taken by several astronomical centers around the world to send probes and we can only encourage this.

In fact, with its fluid composition made of plasma, the sun constitutes for us the natural laboratory where we can understand the phenomena taking place in space. It is a colored vacuum which, with its rotation, helps us to understand certain phenomena that take place in the Universe. Inside the Sun, if we can penetrate it, we will see all the phenomena caused by the energies on earth and in the Universe.

We know that, for low energy chemical reactions, the study is done from atoms as if they were the smallest constituents of material. But for thermonuclear reactions such as those that make the Sun shine, the atoms are used together with neutrinos, which are considered to be elementary particles. This means that even though the Sun is considered by some to be a second– or third-generation star, the study of this inner fluid can also lead us to understand what really exploded during the Big-bang, described by Moses in the Bible as "the separation of the waters from the waters".

Therefore, being relatively close to us compared to other stars, the Sun provides us with a unique opportunity to improve our knowledge of the stars while giving us insights into the nature of the Universe.

1.2. The source of sunlight

The Sun shines because of the nuclear fusion reactions taking place inside it, while releasing energy. As a result of gravity, the hydrogen nuclei within it compress strongly as they fuse and produce helium, further contributing to the release of energy. In other words, the Sun transforms mass into energy, through nuclear fusion reaction. In its core, which is the seat of nucleosynthesis, nearly 620 tons of 620 million tons of hydrogen are transformed each second into 615 million tons of helium. The difference in mass during this process is converted into energy that radiates outwards. This is how the Sun shines and it is the same process that makes all stars shine as well

The sun is a ball of gas, but its gravity is so strong that it appears, seen from a distance, as a solid body. In fact, it is a boiling star of a semi-gas, semi-liquid substance called plasma. Why doesn't it spread out like a river ? It is because it has a core at its center where all material is compressed by the effect of gravitational force. This phenomenon causes so strong pressures that they raise the temperature at its core to more than 15 million degrees. In the scientific documentary "All the Planets of the Solar System", this temperature takes a million years to rise to the surface, where it emerges with a heat of 6,000 degrees.

The heat produced in the process is evacuated in two ways : by radiation, as in sunlight, or by convection movements manifested by protrusions or solar wind emissions.

While it takes only eight minutes for sunlight to reach us, heating and lighting to the earth, some material and gases released by its interact with the surrounding material and gases attracted by its gravity, creating a kind of greenhouse effect like carbon dioxide in the earth's atmosphere. By blocking out the heat like a lid, the heat sometimes rises to more than 2 000 000 degrees Celsius in its outer atmosphere. Science talks about the corona of gases enveloping the Sun.

Having reached this very high temperature, the energy causes the new atoms to burst, causing more heat and heated gases to be released and ejected as plasma into space.

Let's look at the internal structure of the Sun in this image and understand what the collision of fields can create in an empty Universe :

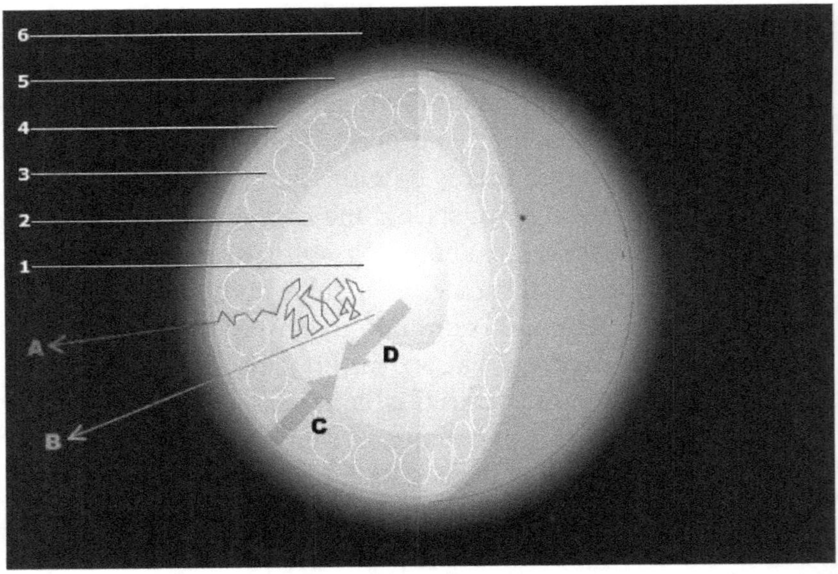

Structure of the sun

Keywords :

1 – Heart of the star, seat of nucleosynthesis (from the center to 0.25 solar radius).

2 – Radiation zone (between 0.25 and 0.7 solar radius).

3 – Convective zone (from 0.7 solar ray to the visible surface).

4 – The photosphere (about 400 km thick)

5 – The chromosphere (about 2,000 km)

6 – The solar corona visible during total eclipses

A – Path of a photon from the nucleus to space, very slowed down in the radiation zone due to the density of material.

B – Path of a neutrino which, interacting very little with material, can cross the star in a straight line.

C – Compression due to gravity.

D – Dilation due to the energy produced in the heart of the star.

(Source : http://www.elementschimiques.fr/ ?en/decouverte/orig -de-la-matiere).

Section 2
Lessons from the Sun

2.1. Gravity creates boundaries

Therefore, just like the Sun and other stars, which are gaseous and semi-solid planets, their boundaries are porous, i.e. permeable. Their boundaries are shaped by gravity. The temperature difference between the core and the surface causes the upper layer to acquire a thick mass, forming the boundary. The same applies to the stars of the entire Universe. This is the first lesson we learn in the study of the Sun.

2.2. The laws of conservation

This lesson allowed me to answer another very important question, namely how does astronomical calculation fall exactly as if we were in a closed Universe ? The answer comes out clearly. We have argued that the Universe is dynamic and self-rotating. This rotation creates gravity, forming a superficial boundary, invisible to the eye or to devices, where galaxies move, so that its expansion does not affect its future. Its consequence is the outcome of the laws of conservation.

Conservation laws are a set of physical laws according to which certain measurable quantities remain constant. Hence the principles of isotropy and homogeneity, which are conservation laws derived from the symmetrical shape of our Universe. The principle of space homogeneity states that the laws of physics are the same at any point in space. This is the law of conservation of the quantity of motion. The principle of isotropy of space means that all directions in space are equivalent for the laws of physics. This is the law of conservation of the cinetic moment.

It was the French chemist Antoine de Lavoisier who formulated for the first time in the 18th century a law on the conservation of material or mass, which was articulated as follows : "Nothing is lost, nothing is created, everything is transformed". This means that in a chemical reaction, the total amount of material of the compounds in a closed system remains constant.

This law was followed by the law of energy conservation, according to a closed system with conserved total energy. It was issued in the 19th century by the scientists Hermann Von Helmholtz, Julius Robert Von Mayer and James Prescott Joule. This law is the first principle of thermodynamics. It was enriched by Albert Einstein in 1905. In his theory of relativity, he demonstrated that there is a mathematical equivalence between mass and energy. Hence his famous formula of E = mc2. This is the law of conservation of the mass-energy set.

Finally, we have the law of conservation of electric charge which stipulates that the sum of the electric charges carried by the particles in a closed system remains unchanged before and after transformation. Several similar laws also apply in particle physics : conservation of color charge, low charge, hypercharge, isospin, baryonic number and leptonic number. (Cfr : "conservation, laws of."). Microsoft® Encarta® 2009 [DVD]. Microsoft Corporation, 2008).

All these laws of conservation are part of the second lesson obtained from the Sun on its shape, created by gravity and which we project on the Universe. My hypothesis is that the astronomical calculation is correct because there is something that encloses us. I was surprised to find that some astrophysicists also believe that our Milky Way galaxy is enclosed by a kind of bubble, which indeed agrees with my conclusion. (Cfr : sciencesetavenir.fr/space/astrophysics/hubble-bubble-our-galaxy -is-is-closed-in-a-bubble-explanations-by Olivier Lascar, on 03.01.2017).

Some astrophysicists believe that our Galaxy is enclosed by a kind of bubble.

Hubble Bubble : what if our galaxy was enclosed in a bubble? © ESA/HUBBLE &NASA, ACKNOWLEDGEMENT : JUDY SCHMIDT

(Sciencesetavenir.fr/espace/astrophysique/hubble-bubble-notre-galaxie-est-elle-enfermee ?).

2.3. Is the Universe finite or infinite?

Without gravity, the sun can explode and unfold like lava. Thus, it is through its gravitational force as plasma liquid that it has created a boundary. The Solar System itself, which is made up of all the stars that are under the influence of the sun's gravitation, also has a porous boundary called the heliosphere.

These considerations lead us to make our choice on the concerns of the French Astrophysicist J.-P. Luminet in his documentary conference "the shape of space". He proposed three hypotheses :

- The Universe is infinite;
- The Universe is finite (without limits), but larger than the visible Universe;
- The Universe is finite (boundless), but smaller than the visible Universe.

On the basis of everything we said, we come to the second conclusion, i.e. that the Universe is finite (without limits) but larger than the visible Universe. This is what astrophysicist J.-P. Luminet calls the hypersphere. This is the third lesson we have learned from the sun.

So our choice was dictated on the basis of our study of the Sun and as I said, it is explained in the following way by astrophysicist J.-P. Luminet :

"What is certain is that if the Universe is finite, it has no edge. Otherwise, what would there be beyond the edge, if not the Universe itself» ? (Cfr : futurascience.com/10 questions essentielles sur l'Univers/ le 19/08/2019).

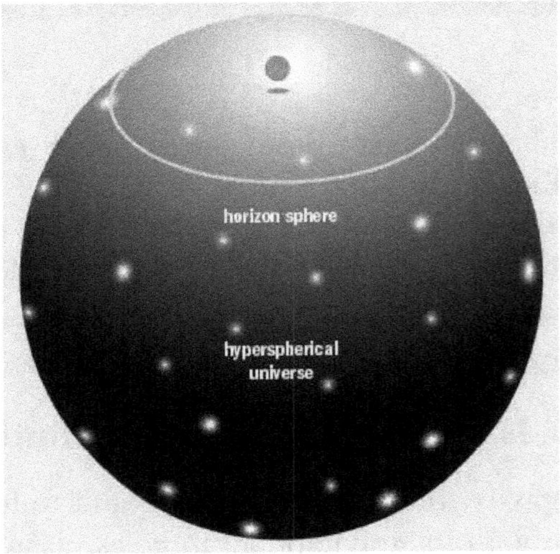

(In a Universe that would be finite but without an edge like a hypersphere, the observable Universe represents only a subset bounded by a cosmological horizon beyond which we cannot see. Public domain). (Cf : futura-sciences.com/buildsv6/images/mediumoriginal/d/e/7/de7da23f61_65821_Universs-jeanpierre-luminet.jpg).

2.4. Cavities in the Sun

I also talked about the black holes created by the rotation of the Universe because they are generated by the difference in towers and tensions at different scales. And then I said, to help my readers understand, that black holes are like tornadoes on earth. The study of the Sun shows us that these cavities also form inside the Sun, because of the same phenomenon and tornadoes are bigger than the earth.

Artist's view of the solar magnetic field. Werner Heil, public domain

(futurasciences.com/buildsv6/images/mediumoriginal/c/f/7/cf737 b45f6_65457_soleil2fs.gif)

Indeed, the solid body does not dislocate when they rotate. The difference in the speed of their rotation at different altitudes is due to their shape. But, fluid bodies, liquid or gaseous, although held together by gravity, their rotation is like a stream of water at different scales. This creates friction and generates magnetic fields. And in these fields, bubbles of gas can be formed, generating phenomena described above, namely tornadoes or vortices inside. It is these gas bubbles that are at the thrust of the lava which are observed during volcanic eruptions. This is confirmed by the observations in this sentence :

*"Moreover, the interior of the Sun is in constant agitation : gas bubbles are born at a depth of 200,000 kilometers and rise to the surface; when they burst, it's as if we were listening to the exit of organ pipes». (*Cfr : futurascience.com/10 essential questions about the Universe/ on 19/08/2019).

The space vacuum, behaves like a fluid or plasma, and generates black holes, known as mini-black holes. This is the fourth lesson we can learn from the Sun.

2.5. Electrical and other phenomena in the Sun

The sun also produces volcanic eruptions in the form of solar storms. It is a projection of fire, causing the ejection of coronal mass and solar eruptions.

Earth waves are also found in the Sun's material and it's called plasma ejection. And inside the Sun, we find all the natural electrical phenomena of the Universe such as lightning and flashes of lightning, etc... Astrophysicists who have knowledge of many of the mysterious phenomena of the Universe find here a star that can meet their expectations.

Composed of fluids, the Sun has within it the behavior of what is visible as a liquid and what is invisible as wind (gas). This is how Moses described the original material of water, i.e. as a fluid.

Thus, to unravel the mystery of the Sun is to enter the natural laboratory of the Universe.

In conclusion, all the lessons learnt about the study of the Sun are for us part of the causes that explain all the effects we have just described. And this starts with the laws of conservation to how a vortex zone can be created in space, and also discovers how space became a laboratory where magnetic fields are created.

Chapter VIII
The physics of creation

Section 1
The principle of creation

The Principle of Creation is a branch of the teaching in the book of the Divine Principle by Rev. Dr. Sun Myung Moon. It is "the only instruction manual of the Universe" for mankind. From the time of creation until the advent of Rev. Dr. Sun Myung Moon, humans have been without a true parent who could teach them why the Universe was created, what was God's principle and why this principle was not achieved. Apart from the Principle of Creation, which is exploited only in this series, we find in the same book of the Divine Principle, the origin of man's falls and God's providence for its restoration. We will exploit certain points in other books.

So, in order to know the Universe and these principles better and to be able to explain it better to scholars without using parables and symbolism, but an appropriate scientific language, Rev. Dr. Sun Myung Moon was obliged to study engineering in Japan. These conclusions, which explain God, the origin of the Universe and its eternity much better, are recorded in the Principle of Creation, a chapter of the Divine Principle, as well as in his many works. And as I said, I have come to conceive the Universal Cosmological Model on the basis of these works.

As the Bible and the Koran, the Principle of Creation, a branch of the Divine Principle is considered by the followers of Rev. Dr. Sun Myung Moon as a revelation from God to their master. We should respect their point of view and also offer the same respect to their message as to the Bible and the Quran. Thus, rather than talking about the Principle of Creation which has a religious connotation, we have selected those points that are of interest to theoretical physics and have called it the Physics of Creation. The latter also encompasses all the theories that attempt to unravel the mysteries that go from nothingness, emptiness, to material.

In Rev. Dr. Sun Myung Moon's Principle of Creation, we have sorted out four points that will be part of the "Physics of Creation", just like quantum physics. Through these four points of the Creation Principle, many of the riddles of the Universe will be solved by scientists, as Dr. Moon said in one recent speeches said :

"If you take what is taught in the Divine Principle and start to apply it in the different fields, you will discover and be able to explain many things (mysteries)".

It concerns the following points :

– The dual characteristics of God, which explains the functioning of the Universe;

– The Universal Prime Energy : This describes the source of energy and its extension in the Universe;

– The period of growth of created beings, this reveals that the biblical days are periods;

– The Spiritual World and the Physical World, of which Rev. Dr. Sun Myung was the only one to describe the links between them and creating the cosmos.

All these contents of the Principle of Creation are synthesized in Volume I and developed in the book of the Divine Principle. I myself bear the responsibility for the judgments of the selected writings or verses. Thus, I offer these two definitions :

– The Universal Cosmological Model is the scientific restitution of the biblical creation account. Its scope covers from the moment the biblical creation narrative speaks of a "formless and empty earth" meaning the Universe from its origin to the last stage, which is the creation of man and woman.

– The Physics of Creation, encompasses all the branches of physics or all the thoughts of physicists, religious or philosophers who thought on the path from the vacuum to the first particles.

– The series "The Verses of the Scientists" extends its field of action from an empty Universe to the second advent of Christ.

Section 2
Celestial Mathematics

Celestial mathematics comes from the culmination of our understanding of the Principles of Creation, a branch of the book of the Divine Principle by Rev. Dr. Sun Myung Moon. It is part of the Physics of Creation, a branch of physics created by me, which encompasses all the physical branches or processes taking place at the time when the earth was still formless and empty according to the Bible, or the early universe in scientific terms. These equations are solved differently from mathematical equations because here it is the entities and it is what is invisible in them that are put together.

For example :

– In normal mathematical equations : 1-1 = 0 or zero

– In celestial mathematics, 1-1 is inappropriate, because entities do not contain the same number of positive and negative elements within them. On the other hand, their subtraction creates a dysfunction and not annihilation.

Thus, in a Christian marital union, 1 + 1 must result in 1.

Since this is an equation that requires a share of responsibility on the part of the spouses, it reads : delta 1 + delta 1 = 1.

In this equation, we aim at what is invisible to the naked eye, such as characters. The spouses must strive to unite their characters so that this does not become the cause of separation. On the contrary, they must harmonize all the time to keep 1 as an answer.

Hence the formula a + b = ab instead of a+b; which means 1+1=1.

The reference verse is : " This is why a man leaves his father and mother and bonds with his wife, and they shall become one flesh" (Gen. 2:24).

– In Christianity we speak of God the Father, God the Son and God the Holy Spirit. The addition gives 1 + 1 + 1 = 1.

The answer 1 represents what is inside them, namely Christ, hence the verses "*whoever has seen me has seen the Father*" (Jn

14:9), *"I and the Father are one"* (Jn 10:30), *" But the Counselor, the Holy Spirit, whom the Father will send in my name, will teach you all things and remind you of everything I have told you"* (Jn 14:26).

Even though their total adds up to 1, each one is an individual. Hence the biblical verses : "Jesus is seated at the right hand of the Father" (Act 7:56), "but that the world may know that I love the Father, and that I am doing as the Father has commanded me, stand up and let us depart from here" (Jn 14:31), " But the Counselor, the Holy Spirit, whom the Father will send in my name, will teach you all things and remind you of everything I have told you " (Jn 14:26). This is to say that people should not expect a Counselor with the face of Jesus or the Holy Spirit who inspires pastors.

2.1. The image of God

The philosophy of celestial mathematics gives us the image of God matched by the following formula :

God = Unity x Love

(Di = Un x Am)

Where there is unity with true constructive and noble love that is where God is.

Love = God/Unity

According to scientific mathematics, it reads : God divided by Unity, but in celestial mathematics it reads "Unity centered on God".

Unity = God/Love

According to mathematics it reads God divided by Love, but in heavenly mathematics it reads "God-centered love".

2.2. The outer form of God

We have :

Divinity = In [(Ci .M) + (Po.Ne) + (Ms. Fe)]

In : Invisible

Ci : Interior feature

M : mass

Po : positivity

Ne : negativity
Ms : Masculinity
Fe : femininity

Therefore, all external forms and their attributes are contained in God, but in a harmonious way. Harmony can only be created by a perfect turn. A perfect turn is one in which the speed between two points is achieved almost simultaneously. This speed is the **In**. This is matched by the biblical phrase "*God is everywhere and at the same time*" (Mt 6:18).

To be everywhere and at the same instant in the Universe, one must rotate at the speed of the dimension of the cosmos per millisecond. This is the variant value of **In**, since we do not know the limits of the dimension of the cosmos.

What is the equivalence between the speed of **In** and the speed of E=mc2 in Albert Einstein's equation ?

First of all, my understanding of E=mc2 differs from physicists. Reading the different textbooks, I have discovered that the understanding of this formula has become like a biblical verse, i.e. even if one knows the components, there are many interpretations for its understanding by theoretical physicists.

In the Universal Cosmological Model, we have adopted this formula as the transforming force that begins in **In**, passing through the virtual energies in an empty Universe (the earth was formless and empty) to the completion of the mass. It is this last part which concerns the E=mc2 which is considered as the absolute limit in science.

For an illustration, we said in the introductory part that in order to evaluate large distances; astronomers have used the reference speed of light in a vacuum, which is about 300,000 km/s, or 7 times around the earth in a single second. This is how the appearance of light always seems spontaneous to us. According to the limited relativity we have said, this speed is a limit that cannot be exceeded by any object with mass.

So, if we planted a stake without mass in the center of the earth at the equator, with a needle of 6,380 km (which is half the equatorial diameter of the earth), and we operated it at the speed of light, after one second it will pass 7 times over each nation. Now, in the formula "Divinity=In [(Ci.M) + (Po.Ne) + (Ms.Fe)]", it is the variant value of **In** that corresponds to the simultaneous speed to reach the furthest point in the Universe from the earth.

The furthest galaxy from earth discovered up to the time I was writing this book is called GN-z11. According to the calculation made by astrophysicists when it was discovered in March 2016, its light travelled 13.4 billion years before reaching us. Its actual distance is estimated at 32 billion light years (al) from the Earth. (Cfr : Wikipedia : GN-z11).

However, 1 al = 9.461 billion kilometer. Thus, this galaxy is 32 billion x 9,461 billion kilometer away from the Earth, which results in 302,752 x 10^{18} km.

So, for the moment, i.e. while we are writing this book, the variant value of "**In**", which also represents the speed at which God as an entity rotates to control the entire observable Universe, is 302,752 x 10^{18} km/s. This speed makes God omnipresent and permanent over the whole universe.

– If the speed of 300,000 km/s represents 7 times the circumference of the earth and if we take it as a reference, the speed of 302,752x10^{18} km/s will represent 302,752 x 10^{18} km/s x 7/300,000 km/s, which will give approximately 7,064x10^{15} times the circumference of the earth per second.

This is how God is more than present in us and nothing escapes Him, as Jesus Christ confirmed in these verses :

Mt 10:29-30 "Are not two sparrows sold for a penny? Yet not one of them falls to the ground without your Father's consent. But even the hairs of your head have all been counted".

Between 300,000 km/s of the absolute speed of light to arrive at the result of e=mc2 and 302,752 x10^{18}km/s to arrive at Dv = In [(Ci.M) + (Po.Ne) + (Ms.Fe)], we can see the travelled road to move from the spiritual universe to an empty Universe. At the beginning, I myself didn't know that there was a great chasm between the spiritual and the quantum vacuum.

Up to now, science has not even been able to reach the speed of 300,000 km/s in the particle acceleration laboratory to understand the visible Universe. So will it manage to reach 302,752 x 10^{18} km/s in the century to begin to understand the transition from spiritual to physical phenomena? What are the different stages that have been crossed? Moreover, the variant value of in is evolving as man's vision progresses to discover the frontiers of the Universe.

We have said that everything in the Universe was generated by a movement of 302,752 x 10^{18} km/s according to the verse "A wind of God moved over the waters" (Ge 1:2). Scientifically speaking, this movement was later characterized by the collision of particles, annihilation and fusion releasing more and more energy, creating particles and anti-particles.

Indeed, scientists have discovered that every particle is associated with an antiparticle of the same mass, but of opposite electric charge. When a particle collides with its antiparticle, both masses are immediately transformed into energy that materializes into other particles and antiparticles. The collision cycles continue while creating matter and anti-matter. And in the end matter won.

As we said, the mass of the particle and its antiparticle, of the matter and its antimatter are not only mathematical values that can annihilate each other, but they are entities of which we don't know how they were formed. Thus, in their collision, they always created something different that in the end the matter prevails on the antimatter. This explains what we said at the beginning– the equations of the entities are solved differently from the mathematical equations because, it is what is invisible in them that is put together. In celestial mathematics, 1-1 is inadequate, because the entities do not contain the same number of positive and negative elements within them. On the other hand, their subtraction creates a dysfunction, not an annihilation.

At present science is still trying to understand how particles prevailed over antiparticles and how material prevailed over antimaterial. The ideal of God or the will of God does not exclude scientific explanations, and it is up to man to discover it. As for us, we have already brought our literal version in the point concerning the dual characteristics of God to Volume I.

The collision is always perpetual, caused by the inherent properties of material according to the dual characteristics of God matched in the formula :

Divinity = In [(Ci.M) + (Po.Ne) + (Ms.Fe)], led to the

energetic formation that created the first deformation of the fabric of space-time, described as the abyss in the biblical scheme of creation in Ge 1:2 (there was darkness on the surface of the abyss).

The confinement of the original material, which was becoming more and more dense in its perpetual movements, biblically described as "a wind of God moving over the waters", caused more and more warming. The heat released illuminated the pre-big-bang universe for the first time, accompanied by the biblical verse "And God said that there was light, and there was light" (Gen 1:3). As the density became greater and greater, it even prevented light from escaping, and created what science calls a deformation of the fabric of space-time (biblical abyss).

Proponents of string theory speak of a star that collapsed into a Pre-Big-bang, three-dimensional Universe.

"*Indeed, for Albert Einstein in general relativity, "any source of energy leads to a distortion of space-time*" (Cf : Ir. MPANZU KIADI BABAKA and Ir. NIMI M., Practical work : explanation of quantum mechanics, UPN -RDC– 2017, unpublished).

According to general relativity, space is represented not as a gas or a vacuum but rather as a stretched rubber cloth. Thus, objects with masses are like balls placed on rubber while deforming it.

According to general relativity, space is represented not as a gas or a vacuum but rather as a stretched rubber cloth. Thus, objects with masses are like balls placed on this rubber while deforming it.

Nevertheless, part of their theory is in line with the biblical scheme of creation, scientifically reconstructed in the Universal Cosmological Model, namely the formation of a mass (biblical water) which will turn, seen from afar, into a kind of star or a fire (and, light was), digging an abyss in the fabric of space-time called, by the proponents of a Universe before the Big-bang, a black hole.

Black holes, which we will see later, are like wells of space-time where everything that comes close to them falls into them and cannot come out. Out of ignorance of biblical language, John Wheeler in 1967 called them black holes, even though it is the word abyss that best suits his definition in biblical glossaries, namely : *"a mysterious World of the primitive ocean, a place of darkness and evil powers, of distress and perdition."*

Did the Universe come out of the black hole? For some scientists, when a black hole is full, it will either expel its contents or vomit them into a wormhole, and for others, the Universe, being condensed to the extreme, has undergone a sudden expansion called the Big-bang or "the separation of the waters from the waters" according to the biblical scheme of creation.

What everyone has in common is the recognition of the Big-bang phenomenon by science and the separation of the waters from the waters in the Bible. The details are described in Volume I.

2.3. The inner power of God

God's inner strength is the source of his eternity :

God = First Universal Energy

The dynamism of the energy within God creates within it the Universal Prime Force. The Universal Prime Force is the fundamental force of God, the Creator. It is also the fundamental force of creation, the force with which God endows each individual being or body with truth when it is created as the substantial object of God.

Thus, God is the origin and source of all energy. By coming from God, energy already possesses the divine inner characters.

This is supported by this sentence from the Principle of Creation :

"According to modern science, all particles that make up atoms are made of energy. In order for energy to form particles, it must have an inner nature that leads it to take on specific forms too". (Cf : Presentation of the divine principle : ed. Culture et Paix, Paris, 2013, Pp 21-26).

So, the great scientific enigma is not God who is a real entity, but the passage from the invisible (spiritual) or nothingness to virtual energies in an empty Universe.

We said it at the beginning, that there is a speed greater than the speed of light, it is the speed of thought. That is why I said that for astrophysicists this book should not be read as a novel, because it is about situations which we shall bring out mathematical bases.

Albert Einstein's formula of E=mc2 (energy = mc squared) allows us to have energy values. This formula already explains the behavior of the elements in a formless and empty Universe. But in order to get to this point, we have the original basic formula :

Divinity = In [(Ci .M) + (Po.Ne) + (Ms.Fe)],

This formula explains the addition of God's dual characteristics. By inverting each dual characteristic, we get the answer that each pair is first characterized as energy based on the original Universal First Force which is equal to the speed of radiation, the latter being equal to the speed of thought.

The slightest decrease in this speed allows the transformation from invisible to visible energy (virtual energy), or what esoteric religions call charge. Thus, in the assertion : God = Universal Primary Energy (UPE), this force was transmitted within the entities by the phrase "God breathed" which man uses figuratively. Thus, within the entities, it manifests itself in the following way :

Di = UPE

UPE = In(FPU)

In : Invisibility

UPF : Universal Prime Force

In each entity or being, we find the energy of God called the Universal prime energy. The Projection of these entities leads them to unite according to their origin and thus generates between them an action of giving and receiving. This generates the Universal prime force. The Universal prime force, generated by the internal action of giving and receiving allows the entity to exist and be active, while creating diversification.

All this is only a part of the fundamental formulas, it is up to mathematicians and physicists to develop and enrich it after having understood the Principle of Creation, a branch of the chapter of the Divine Principle's book Rev. Dr. Sun Myung Moon.

2.4. The determination of times in the Pre-Big-bang period

As we have said it, the ultimate point marking the separation between the Pre-Big-Bang and post-Big-Bang Universe was the Big-Bang. To calculate the duration of the Big-bang times to the present day, astrophysicists have based their calculations on the distance that light has travelled from the farthest star to us, taking the speed of light as the unit. The result was that the Big-bang or "the separation of the waters from the waters" in biblical terms took place at about 14 billion years. This result seems to be confirmed to these days by the study of fossil radiation.

Thus, a visible reference was used to calculate the duration of the observable Universe. But what about the duration of the time before the Big-bang, whose existence science has not yet approved?

Here, the furthest point in the universe from the pre-Big-bang to the extremely condensed Universe is the Big-bang. For the pre-Big-bang Universe, the Big-bang is the final moment. Since there is no point of reference in the pre-Big-bang beginning according to what is written "*in the beginning the earth was formless and empty*" (Ge 1:1-2), i.e. a Universe where there was only emptiness or energy, we will base our calculations on the last phenomenon that science has discovered, namely the Big-bang testifying the existence of the condensed Universe.

According to science, the suspected temperature of the condensed Universe was close to 10^{15} degrees. This temperature was reached, we believe, during the cosmic deflation that began when the Universe was still formless and empty, until it condensed or caused these energies to condense. Thus, to calculate the duration that this time took in the Universal Cosmological Model, I have based myself on the following reasoning, which astrophysicists have to put in the formulas.

Our unknown, X, is the zero time characterized by an initial temperature. The first known value, Y, is the final temperature which is about 1015 degrees. The third known value is delta Z, which is the speed at which deflation took place to condense the Universe, bringing it to reach the maximum temperature of about

10^{15} degrees.

In a Pre-Big-bang Universe that was formless and empty, as we have said, things happened with a decreasing speed starting with the divine speed of 302.752 x 10^{18} km/s. This speed is taken from our formula below representing the **In** :

Divinity = In [(Ci .M) + (Po.Ne) + (Ms.Fe)].

In : Invisible

Ci : Interior feature

M : mass

Po : positivity

Ne : negativity

Ms : Masculinity

Fe : femininity

So to solve our equation, we have the following elements :

Y = 10^{15} degrees

ΔZ = 302.752 x 10^{18} km/s

X =? Where time of the first energy collisions in a formless and empty Universe.

Moses' answer to this equation is 1, i.e. it took a biblical day from X to Y (Ge 1:5). That is, a biblical day has nothing to do with our earthly days (2P 3:8). Astrophysicists know that it is not enough to just take the calculator and point out the answer. Several factors have to be taken into account, for example, in cosmic inflation there has been a reduction in heat caused by an expanding Universe. Here, knowing Y which is approximately the maximum value, there is a reduction in heat as we move back from the condensed Universe to a deflating Universe.

In the Bible it is written "In the beginning God created the heavens and the earth" (Gen 1:1). Indeed, the cosmos has gone through several cycles. In the beginning, there was only the celestial sphere called the heavens. After several cycles, through the divine projection, characterized by radiation, some of the spiritual energies began to density there, creating the sphere called by Moses "a formless and empty earth" or the Universe. The first collision or fusion of these energies made it still dense creating the first deformation of the fabric of space-time or the axis on which everything is poured. Moses described this as the movement of a wind from God was moving over the waters into the abyss. So the answer to the above equation is not about the beginning of the Universe but about the beginning of the first collisions in the period before the Big Bang.

Until then, science had been complicating the Bible, but who knew that one day the situation would turn against scientists. The biblical scheme of creation, once rejected by science, and scientifically reconstructed in the Universal Cosmological Model became the targets to be reached by scientists for the understanding of the Universe.

For the Italian scientist Gabriele Veneziano, this time could not be defined because the universe comes to us from an infinite void.

Section 3
The quantum world

The quantum world or the quantum Universe is the Universe of the infinitely small, not accessible to common mortals, but which confirms to scientists the attributes of God through the behavioral study of particles. The so-called "quantum" theories describe the behavior of atoms and particles. Scientists have been able to observe that the quantum world (limited to the infinitely small, but which can have repercussions on a larger scale) behaves very differently from the macroscopic environment to which we are accustomed. It is a milli-microscopic world. It is the subject of quantum mechanics, a branch of quantum physics. For the Universal Cosmological Model, as we have said, it is part, together with the points taken from the Principle of Creation, of the physics of creation.

3.1. The dual characteristics of God and quantum mechanics

Quantum mechanics is the branch of physics that studies and describes the fundamental phenomena at work in physical systems, particularly at the atomic and subatomic scale, commonly known as the microscopic scale. Thus, it is the fundamental theory of the particles of material constituting the objects of the Universe and of the fields of force animating these objects. (Cfr : futura-sciences.com/sciences/definitions/physical-physical -quantics-13197).

The American physicist, Richard Feynman, said that : "If *you think you understand quantum mechanics, it is because you do not understand it*". And he did not hesitate to call anyone who claimed to have mastered this science an idiot. For scientists, quantum mechanics is full of mysteries, surprises and paradoxes that force them to rethink the way we conceive material and even physics in general. This is because the source of everything,

"God", has been neglected by science.

In fact, several phenomena studied and revealed by quantum mechanics do not belong to the post-Big-bang era ; and I think so. Quantum mechanics is in fact the gateway to the transformation from the spiritual to the physical, that is to say, as soon as the word earth was first pronounced in the Bible through the verse "*In the beginning God created the heavens and the earth*" (Gen 1:1), *an earth that was still formless and empty*, that is to say the Universe (Gen 1:2) was born. It was at that moment that the aspects of God, made material in the Universe, began to return to unity according to the aspect of the Creator, thus creating the cosmic deflation that characterized the Pre-Big Bang Universe.

Therefore, the phenomena studied in quantum mechanics, made possible by a deflating and condensing Universe, do not fit with the one studied in general relativity characterized by an inflating and expanding Universe.

As we have said it above, the Nobel Prize winner Richard Feynman had the honesty to declare that "All those who claim to understand anything about quantum theory – including physicists – are liars".

The scientific commentary of this quote is as follows :

"*This means, in plain language, that the internal mechanisms that govern the behavior of particles are still completely unknown. Even if the statistical approach of quantum physics allows extremely accurate predictions, it is still only an external and clever way of describing things». (*Cfr : zevengeur.wordpress.com/la-chromo-dynamic-quantum).

We have stated that these internal mechanisms are described in the Principle of the Creation of Rev. Dr. Sun Myung Moon, but not yet integrated into science because it is housed in a religious book given its status. The books in the series "The Verses of the Scholars" are the first to bridge the gap between religious and scientific teachings and will henceforth allow one to enter knowledge and the other to leave ignorance.

The Rev. Dr. Sun Myung Moon who was the first to have brought out its physical characteristics in the Principle of Creation, a branch of the book of the Divine Principle, made it possible to understand the mysteries of the Universe, at least for the scholars who will master the explanations. In its movement, the Principle of Creation is taught to all members, giving them a broader understanding of the Universe and the spiritual world than followers of other churches. This responds to the messages of Jesus Christ : "*If the guests do not come, call out to the passers-by so that they may enjoy the feast*" (Mt 22:8-10).

I am only a passer-by ; the guests are the scholars in all fields. As I said it, to introduce great knowledge into a sacred book is to hide a treasure from mankind. Myself as a passer-by and a member of the Unification Church, I have been meditating on this Principle for more than thirty years, but I have not yet acquired its full understanding and that is why I am not yet able to express it perfectly. Theoretical physicists have a head start in its understanding compared to theologians and philosophers.

Through the teaching of Rev. Dr. Sun Myung Moon, we understand that the more humans advance in the understanding of truth, the more it allows God to open the vase. On several occasions Rev. Dr. Sun Myung Moon has gathered the scholars by distributing the book of the Divine Principle, which contains the Principle of Creation. But they had never understood his true intention. It is like the more than five thousand people whom Jesus fed with bread (Mt 14:21) ; they only followed him because of bread (Jn 6:26) instead of understanding his true mission.

With the publication of the books in the series "The Verses of the Scholars", all the so-called sacred books containing divine wisdom will now be carefully read even by atheist scholars as a research document, since faith only comes into play later after the discovery of the riches of The verses of the Scholars.

Returning to our subject, God being invisible, in physics, his mathematical characteristics have been matched by the thinkers of quantum mechanics without knowing it. This can be summed up in several points defined by D. Louapre for the 7 wonders of quantum mechanics. *(Cf : sciencetonante.wordpress.com/2013/09/30/the 7 wonders of quantum mechanics).*

Everything that is studied in quantum mechanics finds its cause in the extension of the omnipresence of God ; matched by the formula :

Divinity = In [(Ci.M) + (Po.Ne) + (Ms.Fe)].

In = 302,752 x 10^{18} km/s : initial speed at which everything from God unfolded.

The decomposition of the highlighted elements, taken in isolation, is indeed the subject of discoveries in physics. And their behavior in order to establish theories constitutes the study of quantum mechanics. With all that is written below, I am not adapting religion to physics, but rather the discovery by physical science of what is said in religion.

Let's now look at some of the main points of quantum mechanics that testify the omnipresence of God.

3.1.1. The principle of superimposition

The Bible says that God, being invisible, his perfection, his eternal power and his divinity, can be seen with the naked eye, since the creation of the world, when we consider them in his works (Rom 1:20). In other words, it was in the study of creation that God could be understood mathematically. Thus, in the Principle of Creation, the Rev. Dr. Sun Myung Moon highlighted those characteristics which now make it possible to understand the mysteries of the functioning of the Universe, since it is part of his projection.

We have said that to be simultaneously everywhere in the Universe from the center to the extreme, the divine speed (**In**) is about 302,752 x 10^{18} km/s. At this speed, God is present everywhere and at the same time, and nothing escapes him. It is not like the speed of light of 300,000 km/s which, coming from the sun, takes 8 minutes to reach the earth and takes several billion years if it comes from far away stars. Yet the divine speed simultaneously makes God present everywhere in the Universe.

When we look at things in our daily lives, including living beings, everyone occupies a specific position even if they are in motion. At a given moment, we find ourselves in a well-defined state. This is the so-called "classical" mechanics, i.e. the mechanics of normal objects.

It is not the same with microscopic objects that undergo the dual characteristics of God. Unlike other normal objects, a microscopic particle can be in a mixture of several states. For example, according to observations, when an electron revolves around a proton, it is in all points of its orbit at the same time.

To designate the fact that in quantum mechanics, objects can be in several states at the same time, we speak of the principle of superposition. This reflects one of the divine attributes called "omnipresence", i.e. God has the ability to be everywhere at the same time. He can be active for the one called the living state and passive for the other called the dead state. Since God is invisible, it is rather the person separated from him who finds himself in a dead state while being alive. Hence the Bible verse "let the dead (those who are envious, but separated from God or spiritually dead) bury their death (natural death) (Mt 8:22). This characteristic is recognized in quantum mechanics through the example of Schrödinger's famous cat, this "fictitious" cat which is both dead and alive.

Thus the active presence of God in individuals makes one person living and the other dead; a principle which can be annotated in Astrotheology by the symbolism of :

|Individual> = | Dead > + | Alive >

Indeed, to describe the two states simultaneously, i.e. a person can be both dead to God and alive to the world; physicists have invented two kinds of hooks like this one ".

"And, to superimpose states, they add them together.

To speak a little about Astrotheology, it is not only in quantum mechanics where Schrödinger's cat can be described as being simultaneously dead and alive. Since the universe is still in the grip of evil all creatures find themselves physically in a dead state while being alive (Rom 8:22-23).

To go from one state to another, is described in the Principle of Resurrection, a branch of the book of the Divine Principle by Rev. Dr. Sun Myung Moon. To paraphrase it, the resurrection that people wrongly consider as a physical return to life and as a rising in the clouds, is only the fulfillment of the purpose of creation that every individual is called to achieve is the fulfillment of the purpose of creation as described in the Principle of Creation, contained in the same book.

Let us read the verses mentioned above :

Matthew 4:8-9"*The devil again carried Jesus up to a very high mountain, showed him all the kingdoms of the world and their glory, and said to him, 'I will give you all these things if you will worship and adore me.*

Rom 8:22-23"*Now we know that the whole creation is still sighing and suffering the pains of childbirth. And not only it, but we too, who have the first fruits of the Spirit, are sighing within ourselves, waiting for adoption, for the redemption of our bodies.*

We will come back to this in Volume V for further explanations, this volume is rather reserved for science.

The omnipresence of God, reflected in the principle of the superposition of states in physics, characterizes all the strangeness of the phenomena discovered in quantum mechanics, as we said it in the Universal Cosmological Model, is part of the Physics of Creation, that is to say of a physics before the Big Bang, even if their authors were not aware of it.

3.1.2. The indeterminism of measurement

So far we have been talking about the speed of about 302,752 x 10^{18} km/s of God's energy, for Him to be permanent throughout the Universe. However, this measurement result is potential, since we have referred to the most distant star discovered so far. If one-day science discovers more, it would change the situation but would not alter our formulas in any way. Besides, God is spiritual and we cannot give him physical values with precision; it is the indeterminism of measurement.

Nevertheless, it allows us to understand what was originally the speed of radiation of virtual particles causing energetic collisions, until arriving at real particles as we can read : "*In the beginning, the radiation is so intense that it possesses the energy necessary to materialize all the possible particle/anti-particle pairs, which are immediately annihilated, or destroyed immediately.* "André Brahic in "Sciences of the Earth and the Universe".

This Pre-Big-bang deflation, which offered energy to the virtual particles, ended up generating, due to the confinement of the original material, the first light. These three great Pre-Big-bang stages, which we explained in detail in Volume I, are contained in these biblical verses : *"The earth was formless and empty... A wind of God moving over the waters... And light was"*. (Gen 1:2-3).

This initial speed of deflation, which was degrading as the Universe condensed, rebounded during the Big Bang. This is confirmed by science through observations according to inflation which took place at a speed exceeding that of light. And even some hypotheses claim that the speed of the current expansion of the Universe is higher than the speed of light.

Thus, starting at **302,752 x 10^{18}km/s** during deflation in the Pre-Big-bang period, we can understand why current physics cannot explain the difference in speed between inflation at the beginning and the current expansion of the Universe, judging it too fast as we can read :

"However, in recent years, one problem has particularly bothered physicists : the acceleration of the expansion of the Universe is much faster than it should be".(Cfr : trustmyscience.com/Expansion-University-acceleration-trop-rapid-for-and-be-explained-by-physics, by Thomas Boisson, 26/02/2018).

Returning to our example of the electron that can be present everywhere around the proton, physicists have discovered that every time they calculate its speed, it changes or sometimes repeats itself. This is noted by the sign :

Electron> = |1000 km/h > + |2000 km/h >

This means that by measuring the speed of the electron, one will find either one (1000 km/h) or the other (2000 km/h), but there is no way of knowing in advance which one.

According to D. Louapre in the scientific documentary "the 7 wonders of quantum mechanics", these results have brought about an incredible conceptual revolution in physics. Physicists have long assumed that nature was deterministic : if you do exactly the same experiment twice (in principle), you find the same result twice. But in quantum physics, which is the front door of the transformation of the spiritual into the physical, the results are not the same.

According to the Danish physicist Niels Bohr, there is a fundamental indeterminism that makes the results of measurements depend on chance in a way that is impossible to predict. Albert Einstein did not like his thinking and told him that "God does not play dice", which in our opinion means that everything in creation is assigned a value according to its purpose. As physics cannot question the soul and it can only evoke chance.

For example, a player who shoots the ball into the opponent's goal at 30 meters will not deploy the same energy if he hands it to his goalkeeper at the same distance or passes it to his teammate. By this we mean that there is a goal assigned to everything; and we confirm Albert Einstein's saying that "God does not play dice". In Rev. Dr. Sun Myung Moon's Principle of Creation, this is developed in the point concerning the action of giving and receiving. (Cfr : The Divine Principle, Level 4, Pp 19-21).

Of course, when a particle finds itself isolated for an experiment, in search of its goal, its results will also be modified in relation to a particle wandering around in nature. This is why we speak of the indeterminism of measurement for quantum elements.

3.1.3. The wave-corpuscle duality

One of the questions we ask ourselves is : how did we move from divine radiation to material? Physicists have thought taking into account the light, because for them it all began with the Big Bang. This is recognized biblically by the phrase "the separation of waters from the waters" which took place on the second day or in the second period, confirming that it was not of the beginning. Of course, there was indeed a very eventful first day, also summarized by the biblical verse "*a wind of God moved over the waters*" (Gen 1:2).

We said that the coefficient of the divine speed that we used in the calculations was related to the probability of finding God at every place in the Universe, known theologically as the omnipresence of God. When transmitted to radiation, this radiation became heavier to generate punctual particles with the dual characteristics of God, from which they emanated, and which would later be transmitted to material. Hence the punctual objects according to David Louapre in the documentary "Quantum mechanics in 7 ideas" behave, at the microscopic level, like waves, i.e. they are everywhere at once.

This idea that waves were already material was introduced at the beginning of the 20th century by several researchers, including the French physicist Louis De Broglie. De Broglie proposed the idea of wave-corpuscle duality : particles can follow circumstances and behave either as particles or as waves. This idea put an end to the nature of light, which is made up of electromagnetic waves and photons.

As I have said it, 20th century physicists took light as a reference, unknowingly applying the biblical verse that says *"what is invisible in God can be recognized in the study of what is visible"* (Rom 1:20). This is why I advocate for Astrotheology, which will bring science into a new step with Rev. Dr. Sun Myung Moon's Principle of Creation as a coefficient reference, because he did not have the mathematical time to develop his ideas and neither did Moses ; God's providence obliges.

These two men of God laid all the foundations for the development of Astrotheology. So, the scientists based themselves on the study of light, but we based ourselves on the origin when we talk about radiation.

3.1.4. The tunnel effect

One of God's attributes is that He has no barriers ; He transcends time and space. Our physical barriers are nothing to Him. He even probes our thoughts and hearts (in the religious sense) as this verse attests :

Mt 6:6 "But when you pray, go into your private room, shut your door, and pray to your Father who is there in secret ; and your Father, who sees in secret, will reward you".

The Universe is a divine projection, this attribute is preserved by certain rays and microscopic elements, which are the subject of study in quantum mechanics. If God is global but to each individual, he becomes a personal God known as the Father.

Mt 23:9 And call no one on earth your Father ; for one is your Father, he who is in heaven.

This characteristic acquired by certain microscopic physical elements that one slice can transcend a barrier and end up on the other side, is called the tunnel effect. There are still certain elements that transcend us every second as if we didn't even exist; this is the case of neutrino. All this still makes God the creator, the progenitor and the projector of everything. Everything in the Universe has at least one of these characteristics as described in Rev. Dr. Sun Myung Moon's Principle of Creation.

The tunnel effect is one of the things that happens in the quantum world, but not in the macroscopic world. Science has used its theory, for example, to make microscopes known as "scanning tunneling microscopes", which make it possible to see and manipulate atoms.

3.1.5. The path integral

Believers sometimes have special places for meditation, but God transcends time and space. He does not follow a specific path to reach us and no barrier can stop him. As stated in the tunnel effect, he has no barriers.

This characteristic acquired by microscopic elements means that in quantum mechanics, when a quantum particle goes from one point to another, it passes through all the possible paths that connect these two points. This is demonstrated by several Feynman diagrams that astrophysicists are more familiar with. This approach is scientifically known as the path integral.

3.1.6. Teleportation

Teleportation is an exchange of information in an almost instantaneous manner without the use of voice, gestures or external tools. They are not limited to thoughts only, but cover all psychic areas such as feelings, pains, apprehensions, etc.

Through quantum mechanics, scientists are beginning to explain, based on the behavior of particles, what was previously part of the psyche. This takes them away from religion and philosophy, but for us it is the opposite, all the more so since

God is the creator and has endowed man with all the capacities since his creation.

Indeed, for the French physicist Claude Sarrailh, "*the elementary particles which are the vectors of action have no problem with time (the past, the future). Before interacting, they are everywhere in space-time and time for them does not pass... Let us remember that we are only made of atoms where electrons revolve around which we don't know before observing them where they are (they have no physical reality); And these electrons have the power to go into the future and to come back instantaneously in the present instant, since before observing them they were in all space-time, without constraint linked to terrestrial time because for them time has not passed*". (Cfr : Claude Sarrailh, Le secret de la vie, edited by LIBRINOVA, Paris 2018, Pp 95-100).

Moreover, scientists observed that the state of entangled atoms (a very complicated notion to explain in a book of literature) changes faster than the time it takes for laser rays to reach an object, i.e. faster than the speed of light. This is justifying for us, from the point of physical life, this characteristic of instantaneous deportation.

Many have already experienced teleportation. For example, being attracted to a person in an assembly makes that person turn to you or notice you.

Often in childhood, as soon as I thought of a song while listening to the radio, it would be played within minutes. This reflects the popular expression "*When you talk about the wolf, you see its tail*", which is said when a person appears when you talk about him or her. In some esoteric sects these teleportation exercises are found. And the followers of these kinds of sects can influence certain events from afar.

One of the golden rules of spinning is that you must not think about the person you are spinning, otherwise he will start to turn to you from time to time and may even feel spun. This is also what happens when you tail a girl to approach her, or if you focus on her in a café. Chances are that she will turn to you and notice you before you can even talk to her.

This is because girls have a very deep intuitive sense about themselves compared to men. I must have experienced myself something I've just described, but at that time I didn't know it was called teleportation. Neither did I know that these phenomena are explained in quantum mechanics.

As quantum mechanics impinges on philosophy and religion, one researcher wrote this :

"One gets the impression that quantum mechanics justifies philosophical theories considered bizarre by most scientists, ranging from radical forms of idealism to mysticism". (Cfr; Jean Bricmon, Quantum mechanics for non-physicists, B-1348 Louvain-la-Neuve,Belgium).

For what concern the notion of time, I said that there are speeds that are higher than the speed of light and quantum mechanics proves me right. Moreover, it also demonstrates that the speed of divine invisibility of 302.752 x 10^{18} km/s is not a decoy and I have demonstrated with simple calculations how this was obtained.

As we can see, all the characteristics discovered in objects studied in quantum mechanics reveal the omnipresence of God through their behavior. From this we draw the following conclusions :

– Quantum reality is different from our reality. At the quantum scale, nature functions according to mysterious rules, with particles reflecting God's attributes. It is a reality that originated Pre– Big-bang or the period before "the separation of the waters from the waters" according to the Bible, when the Universe was in deflation. This reality means that quantum mechanics cannot unite with the general theory of relativity without taking this parameter into account.

– Moreover, the entanglement principle, a branch of quantum mechanics, unknowingly corresponds to physicists the biblical and Koranic writings and the Divine Principle that make God the origin of everything. Indeed, through this principle, scientists have discovered that we are all united and not separated according to appearances and this contradicts Charles Darwin's theory of evolution. As in the Bible, many of the phenomena discovered in the quantum world are still subject to interpretation.

In Romans 1:20, we are told that "*the invisible perfections of God, his eternal power and divinity, are seen with the naked eye, since the creation of the world, when we consider them in his works. They are therefore inexcusable*".

Science, by studying the infinitely small in the branch of physics, has never ceased to discover things that can only confirm the divine characteristics. Therefore, each time a physicist discovers one of the divine attributes through creatures, he imposes his name by putting forward a theory. But there is still much to discover.

Thus, the quantum objects that make the conjunction between God and macroscopic objects keep these characteristics, even if it is decreasing as when we channel an electron and observe it.

3.1.7. Quantum entanglement

Quantum entanglement can be understood as teleportation at the microscopic level. According to quantum mechanics, two particles are entangled when they have a common history, here we can even say that when they come from the same atom. The entanglement means that, whatever the distance between them, their behavior is identical. A change in one of them simultaneously leads to a change in the other.

According to Blandine Pluchet, the experiment observed during the demonstration on entanglement in the year 2000 revisited the principle of causality (which says that cause always precedes effect) because the latter is dependent on the time that elapses (Cfr : Blandine Pluchet, quantum physics for dummies, ed. First, Paris 2018, p. 139).

As we can see, this finding confirms our theory of the speed of divine invisibility of 302,752 x 10^{18} km/s, which makes God exists everywhere and at the same instant. And, as we have also said, quantum phenomena only confirm God's dual characteristics.

Since all men come from a single source, which is God, as it is written, '*Then God said, let us make man in our image, according to our likeness*' (Gen 1:26), it turns out that the quantum entanglement phenomena we call teleportation on a macroscopic level are much more experienced on a daily basis by married couples.

Gen 1:27 "God created man in his own image, he created him in the image of God, he created them male and female".

Through this verse we see that the image of one God is formed by the married couple man and woman and not the single ones, whatever the reason.

Chapter IX
Classical thoughts

Section 1
Background

The terms thought to be classic come from economic history. It was Karl Marx (1818-1883) who invented the classical term by opposing classical economists to vulgar economists. The classics were those who sought to determine the origin of value.

In cosmology, on the other hand, what differentiates the two schools, between the vulgar and the classical, is the introduction of instruments in observation and mathematics which now make it possible to base one's thoughts on them. Thus, it is a demarcation between those who based themselves solely on philosophy until the introduction of observation instruments in the 17th century by Galileo. This now marked the beginning of classical thought, although we associate the Polish astronomer Nicolas Copernicus (1473-1543) who, in the 16th century, put an end to the thought that defined the Universe as a geocentric set in which the central point is the Earth, stationary and without movement, around a certain number of revolving spheres (Cf : https://www.astrofiles.net/astronomie-nicolascopernic).

By thinking according to the solar system, Nicolas Copernicus address the idea of heliocentrism, i.e. the Sun is at the center, the Earth and all the stars revolving around it. It should be noted that, at that time, thinking about the Universe stopped around the stars visible to the naked eye. Therefore, according to science it is this thought, which put an end to geocentrism, that is considered to be the beginning of classical thought.

So, theories have followed one another throughout history in an attempt to explain the creation and evolution of the Universe. They are called cosmological theories. Some theories, which are known to be false today, remained in force for several centuries. It must therefore always be kept in mind that a theory, however appealing it may be, is only one model, only one possibility among others.

It must be said that from the 17th century onwards, astronomers began to ignore philosophical models, relying on instruments. From the Polish astronomer Nicholas Copernicus (1473-1543) to the Italian scientist Galileo, who was the first to use the telescope in 1609, astronomy took a new turn, but without breaking away from the ideas of the Church. The real impetus came in the twentieth century when new disciplines with realistic judgment began to take an interest in cosmology, including mathematics and physics. This gave rise to astrophysics, and those who study it now have a monopoly on the theories about the Universe.

Their theories and judgments giving rise to several branches of classical physics to explain the Universe, that is to say, theories whose cosmological judgment and other descriptions of natural phenomena are no longer based on philosophy, theology or metaphysics, but, theories that could be demonstrated and predictions to replace prophecies.

Among these branches we have quantum physics, which is the general name for a set of physical theories born in the 20th century which, like the theory of relativity, marks a break with what is now called classical physics, which by definition includes the physical theories and principles known in the 19th century. Thus all cosmological thoughts are now validated by quantum physics.

The plurality of thoughts, theories and assumptions has given rise to several cosmological models.

But, for us the only true model on the Universe is the one that comes from the Creator who is God. This statement, which was initially a belief, has become real thanks to the Universal Cosmological Model. An astrophysicist will never create something to sprinkle into the Universe in order to change its natural laws. All that scientists are discovering are the effects of what happened from the moment the movement of the Universe (rotation of the original particles) was initiated in the Pre-Big-bang, described as the first day by the Bible. Hence, for us, religious people also have their say through a more realistic science, combining faith and mathematics, which we call Astrotheology.

In the end, we can say that the Universe did not begin with two or more models. The truth is that the Universe began with only one model; the Biblical Model which is beaten by the foundation of the Universal Cosmological Model.

All the cosmological models of today, and those that will come later, are only in their infancy. When we see a child with macrocephaly, some may say that his mother conceived him by Caesarean section, others will say that the doctor performed a large episiotomy. But who can really say how this child came into the world with just one version? It is only his mother if she is alive; otherwise it is the doctor or the witnesses. And, if the latter are not there, there are the hospital documents.

For a concrete case, one day my mother was asked the question "why is your child always alone"? She replied : "He didn't even want to be born; he spent eleven months in the womb". The others could speculate, but she had her own explanation.

For the cosmos of which GOD is the progenitor, in spite of our separation and break in communication with Him because of our sins, we are nevertheless lucky to have a hospitable document which tells us about his birth and it is accessible to all; it is the Bible. The story of creation contained in the Bible is not written by a prophet who received revelations like us, which can be questioned, but by someone to whom it is written of him :

Ge 33 : 11 "The Lord spoke with Moses face to face, as a man speaks to his friend".

Even Jesus Christ was nailed to the cross for having dared, according to the Jews, to contradict the laws that were left by Moses. Those who want to understand more about Moses need only read the fifth volume of the "Verses of the Scholars" series. So, I was saying that the cosmos was born by only one model which is the story of creation contained in chapters one and two of Genesis in the Bible, written by Moses and commented on by other prophets in other books.

When God appeared to me in 2012, I don't know if it could be called a dream or a vision, but he woke me up to tell me that I am changing the sentence that I had written the same night to "from now on, the biblical model must also be taken into consideration like all the other scientific models". God commanded me to delete this sentence and replace it with "everyone must be inspired by the biblical model of creation". It was a great encouragement for me to see the God of eight billion people in humanity according to the statistics of 2020, appearing personally to give me an injunction, although I have to prove it to the scientists.

The discoveries I was making plunged me into joy and sometimes despair if science has not yet come up with these results. Thus, I take full responsibility for all the imperfections of the ideas contained in this book. However, I know that my readers read to understand the Universe and not to judge me.

Since no one has a monopoly on truth and science, and since God treats us all equally. I, for my part, consider the other models as a basis for enriching and explaining the biblical model. My own contribution is the Universal Cosmological Model.

Thus, we are going to confront several classical thoughts in cosmology and model with the Universal Cosmological Model in this chapter, so that revolutionary astrophysicists can make light. Why revolutionary astrophysicists ? Because for classical astrophysicists and conservative theologians, I am only a heretic.

1.1. The development of classical thought

1.1.1. The Static Universe model

In 1917, Albert Einstein proposed a model of the Universe based on his theory of relativity, known as the Static Model. He demonstrated that gravitation is actually a fictitious force caused by the curve of space-time. Convinced that the Universe can only be static, Einstein then assumed the existence of a repulsive force between galaxies that would counterbalance the gravitational force of attraction. To do this, he introduced a "cosmological constant" into his equations – whose role is to oppose the gravitational attraction that justified the movement of the planets, leading to a static Universe.

1.1.2. Edwin Hubble's discovery and the expansion of the Universe

In 1920, Edwin Hubble was the first to discover that there were other galaxies outside the Milky Way. He noticed that some were moving away from where he had photographed them when their light turned red. He concluded that the galaxies were moving away from each other. Thus, he demonstrated that the Universe is expanding (and therefore not static). This fact was recognized by Albert Einstein, who later confided that his cosmological theory was the biggest mistake of his life. It was also the confirmation of the model of the expanding Universe.

Hubble also noted that the further away a galaxy is, the faster it moves. He concluded and proved that a galaxy's speed of remoteness is proportional to its distance. This important relationship was the origin of the Red Shift Law or Hubble's Law.

1.1.3. The model of the expanding Universe

The model of a static Universe was already not unanimous and was rejected by scholars such as the Dutch astronomer Willem de Sitter in 1917, the Russian mathematician Alexander Friedmann in 1922, and the Belgian abbot Georges Lemaitre in 1927 who professed non-static models of the Universe.

The Belgian physicist Abbe Georges Lemaitre, referring to Hubble's conclusions, without relying on the Bible, suggested that if the Universe is expanding, it is because it began with a point. He thus developed the theory of the primitive atom. According to this theory, galaxies are fragments that have been thrown up by the explosion of that atom, hence they are the expansion of the Universe. However, his theory was not well received.

Indeed, Abbe Georges Lemaitre had already opted for models of the Universe with a finite temporal passage and an initial singularity. Some people accused him of concordism, i.e. of having been influenced by his religious convictions. As we can see, the primitive atom. model followed expansion, describing an extremely hot and dense Universe in its initial step, and filled with radiation. This is reminiscent of the "fiat lux" of Genesis, its detractors said.

Albert Einstein himself reacted negatively to this model by declaring : "no, it is too reminiscent of creation". The famous astrophysicist Arthur Eddington, having abhorred this idea, proposed in 1930 a model where the value of the cosmological constant was specially adjusted to describe a Universe without "primitive atom. followed by expansion", infinite in the past and in the future. (Cf : WWW Futurascience, JP LUMINET, Singularities and zero time).

In fact, when a religious person tells a truth, even a scientific truth, he is always treated as having brought faith into play, as the contempt for science. The Belgian abbot physicist Georges Lemaitre replied to his detractors that his theory had nothing to do with faith and that it was scientific. Fortunately for him, as an astronomer, his theory was noticed and supported by the Russian Friedmann. And, knowing that the Russians had nothing to do with faith, his theory was supported.

Moreover, the sequence of events proved Lemaitre right : the observed abundance of light elements, the discovery of cosmic background radiation and its properties, the ages of stars, the evolution of galaxies... All these results militate in favor of a very hot step of the primordial Universe, and therefore of the Big-bang models which will be retained as a name instead of the primitive atom.ic Universe model followed by expansion.

So it is not surprising that the writings of religious men such as the Rev. Dr. Sun Myung Moon (1920-2012), who set out the natural laws of physics in the "Principles of Creation" based on the Universe, are still unknown to astrophysicists. As he has studied engineering in Japan, if he had put the chapter of the Principle of Creation outside the book of the Divine Principle and called it the Physical Laws of Creation, there would be debating among scholars these days.

For scientists at present, Abbot G. Lemaitre address the theory of the primitive atom. based on Edwin Hubble's observations in 1920 when he found that galaxies were moving away from each other. But, as a religious man and astronomer, he already knew that if God created the Universe, it could only begin with a ball that later burst. But as there was no scientific evidence to support his theory, it remained in his thoughts. When Hubble made his discovery, it was an opportunity for him to bring out his theory of the primitive atom. and defend it with all his energy.

Pope Pius XII wanted to support him through religion, but he refused to do so, saying that there was nothing religiously inspired in his theory and that his thoughts were scientifically founded. But, as the first to talk about the primitive atom., he should also have been the first to discover the Big-bang in the Bible if he had recognized that its source of inspiration came from God.

1.1.4. The model of the Stationary Universe

As the theories on the origins of the Universe became more and more raging, in 1948, the English astronomers Hermann Bondi, Thomas Gold and Fred Hoyle presented another model of the Universe : the model of the stationary Universe. The basis of this model is first of all philosophical, and postulates that the Universe cannot have appeared suddenly. Their model is an extension of Friedmann's "cosmological principle" while introducing the word "perfect". According to Bondi, Gold and Hoyle's "perfect cosmological principle", the Universe looks the same at any time and at any point. In spite of its expansion, it has no beginning and will have no end, because the losses of the elements are compensated by the creation of other elements.

As I have said, this model disputed its place in science with the model of the expanding Universe. The discovery of the expansion of the Universe proved that it was not static. Initially, it was this second hypothesis of a stationary Universe that was the most popular.

One of the reasons for this success is that in this model, called the stationary state theory, the Universe is eternal, because the disappearance of material is compensated for by the creation of new material. Conversely, in the Big-bang hypothesis, the Universe has a finite age if we estimate its rate of expansion. Big-bang type models were therefore in difficulty with such observations. "These difficulties disappeared following a more precise re-evaluation of the rate of expansion of the Universe». (See : WWW.Wikipedia : BIG-BANG);

1.1.5. The standard model

The Standard Model is based on the Big-bang theory. But first, let's come back to the latter.

The Big-bang is a scientific theory explaining the first moments of the Universe. It is the beginning of the expansion of the Universe, about 13.7 billion years ago, when the Universe was extremely dense and hot. This is not the beginning of the Universe as it is often said, because science does not specify what there was before (or even if there was a before). It has become a scientific theory because it is partly confirmed by certain astronomical observations and mathematical calculations. Today, religion through this book in the verse Ge 1 : 6-7 "...God separated the waters from the waters...". Also confirms this Big-bang theory.

Consequently, it becomes a natural phenomenon that at one time characterized the Universe at its beginning. The general concept of the Big-bang, namely that the Universe is expanding and has been denser and warmer in the past, it is attributed to the Russian Alexander Friedman, who proposed it in 1922, five years before Lemaitre according to our research.

The term Big-bang was first used in 1950. In the beginning, we said, it was the English physicist Fred Hoyle who coined the term to mock this new cosmological theory. This physicist believed, like many other scientists of that time, that the Universe had always been the same and professed the idea of a stationary Universe.

The Big-bang theory was confirmed in 1965 after the discovery of fossil radiation. Indeed, according to a study by the Russian-born American scientist George Gamow (among others) in the 1950s, the Universe must have been filled with radiation that should have been emitted during the Big-bang and lost it as it expanded. This radiation is today called the cosmic microwave background, or sometimes fossil radiation whose current temperature is 2.7 Kelvin, according to Gamow's predictions. Fortunately, it was not religious, otherwise no scientist would have believed it, and science would not have launched any research.

Since then, all the calculation of the scientists has concentrated on explaining the Universe from the Big-bang, as well as the phenomena discovered in it. This nowadays forms the Standard Model, a term inspired by the Standard Model in particle physics.

(Cf : www. Wikimedia ; Standard Model).

It is also called standard because it is considered to be sufficiently coherent and stable to be the subject of a consensus among specialists in the field.

1.1.6. The Universal Cosmological Model

I designed this model in 2010 to satisfy my curiosity. My concern was to find out why the biblical scheme of creation was never mentioned in scientific books. Even in the history of cosmology it is not mentioned.

In order to find out why this pattern was rejected by science, I drew the scientific and biblical chronology to compare only the stages of the appearance of things according to the Bible and according to science. The only striking fact at the outset is that all the models recognize that man is the last creature as also classified in the Bible.

But as far as the beginning of the universe was concerned, there was no common ground. Science was exhibiting a name that was completely unknown to me : "the Big-bang", and scientists were even exhibiting it to demonstrate that the Universe is the fruit of coincidence. The term "Big-bang" alludes much more to something accidental, non-programmed, whereas believers attribute all manifestations about the creation of the Universe to divine will.

For two years, I made efforts to better understand this phenomenon and all its properties. I discovered afterwards that the "Big-bang" is a reality and is biblical.

In the Bible it is known by the phrase *"the separation of the waters from the waters"* (Gen 1:6), which took place on the second day of creation. Therefore, biblically, it is only a transitory phenomenon and not the beginning of the Universe. Thus, I conclude that the first day of creation was before the Big-bang. These historical discoveries, which from now on will align the biblical scheme of creation in classical physics, surprised God himself who appeared to me and gave me the injunction to make the world understand this as the only cosmological scheme from which everyone should draw inspiration. Since this declaration of faith could not be accepted by scientists, I could not declare it without being able to demonstrate it. This is what I did in Volume I; cosmology according to biblical and ancient models.

I had once asked a French physicist by e-mail if he could tell me something about the energies that were working before the Big-bang. He replied that the Pre-Big-bang is an "aporia", a word that I had rushed to look up in the Grand Robert dictionary, which defined it as "a rational difficulty that seems to have no way out, an insoluble contradiction". I understood directly that I shouldn't count on anyone to develop my ideas on the Pre-Big-bang if I didn't want to be discouraged in my research.

For 10 years I was researching how to transpose religious language into scientific language in order to convince scholars of the truthfulness of this divine statement. This is what we have been doing throughout Volume I, we have said.

In the course of this research as transcribed in the first volumes, I confirm that it is really God who is the Creator of the Universe. The history of biblical creation hides a great scientific scheme which is reconstructed in this book under a new cosmological model called "the Universal Cosmological Model".

It is this model which now guides the biblical creation story from cosmogony to cosmology and will henceforth be the subject of in-depth study by scientists.

To sum up, the Universal Cosmological Model, which reconstructs the story of biblical creation, recognizes 3 stages for the universe.

1) The universe began with nothingness, chaos and emptiness defined biblically by the phrase "a formless and empty earth" (Ge 1:2). Something that is formless and empty is simply something that is materially non-existent. It is like a colorless and odorless gas in a vacuum or in space.

It was in this universe that a qualified energy of God's spirit began to move over the original qualified material of water until it heated up (And light was) (Gen 1:5). This was during the Pre-Big-bang period.

2) Condensed to the extreme by cosmic deflation, there was an expansion defined as Big-bang by science and separation of

waters from waters by the Bible (Gen 1:6-7). This is the Big-bang.

3) Then came the after Big-bang recognized biblically and scientifically from the creation of the stars to the present day. (Gen 1:9-31).

These are the main lines of the Universal Cosmological Model. Therefore, the Universal Cosmological Model is the only model that has just revealed the starting line of the Universe, passing through the Big-bang to the present day.

Chapter X
The Universal Cosmological Model facing the classic models

Section 1
The standard model

1.1 Introduction of the cosmological standard model

Before tackling this point, we first want to see what we ourselves have understood by the Standard Cosmological Model and this after compiling several scientific writings.

Indeed, the model or Big-bang theory and the theory of the Cosmological Standard Model are so closely linked that I cannot discuss it separately in this book. Astrophysicists will understand it for the following reason evoked by one of them in the following quotation :

"From the beginning there has been an ambiguity, a misunderstanding, between scientists and the general public about the Big-bang model. The general public understands "Big-bang" as a description of the origin of the Universe. Scientists, on the contrary, call "Big-bang" the continuation of the evolutionary stages of the Universe from a hot, dense and highly concentrated state to the present day. This state is close enough to the origin but is not the origin. » (CNRS : La saga du Big-bang, Jean Iliopoulos and Stavros Katsanevas).

As one can see, I am part of the general public and I am no exception to the rule. And from the moment we discover new particles and theories to explain this evolution of the Universe from the Big-bang to the present day, we enter the Cosmological Standard Model. We, the general public, as long as we are not yet in the FAC of Physics, do not need to know the different names of the parts of a needle.

For us the general public, the Big-bang model, explains how the Universe was built and evolved. But, from the 1970s onwards, physicists came up with a probable scenario for its history called the "Standard Model of Cosmology" recently renamed "ΛCDM" for "Lambda Cold Dark Material". This model is based on the theories of general relativity, quantum physics and particle physics. (Cfr : zevengeur.wordpress.com/2014/01/16/histoire-de-l'Univers-partie

-4-le-modle-standard-de-laphysique).

In cosmology, the model ΛCDM (pronounced "Lambda CDM") designates a cosmological model representing a homogeneous and isotropic Universe, whose spatial curve is zero, and which contains dark material and dark energy in addition to ordinary material. The Greek letter Λ is usually the symbol of the cosmological constant, which is the simplest form of dark energy.

The standard model of cosmology has therefore covered almost the entire history of the Universe over a period of 14 billion years, the estimated age of the Universe. The objective of the model is to describe the steps that led to today's Universe.

This model has been developed progressively by going back and forth between experience and theory. Thus, it is an evolutionary model where not only theories but also new elementary particles were added as observations were made. This is in response to all the observed cases. Hence the name of the standard model, which covers all phenomena encountered in nature except gravitation for reasons that even theoretical physicists can explain.

1.2. The view point of the Standard Model on dark energy and dark material

"According to the model ΛCDM, dark energy is presented as a property of space itself, the beginning of which immediately follows the period of inflation as described in the equations of state. The ΛCDM model says nothing about the fundamental physical origin of dark energy, but it represents the energy density of a flat Universe. Observations indicate that it has existed for at least 9 billion years". (Cfr : Wikipedia : History of the Universe).

As for dark material, which is a form of invisible material because it does not emit electromagnetic radiation, it has also been included in the Standard Model under the version of the model ΛCDM.

In some literature, dark material is sometimes referred to as "dark material" and in others as "gloomy material". The difference is that dark material generally refers to ordinary material that is too cold or not bright enough to be detected, whereas gloomy material would probably be made up of electrically neutral particles (different from protons and electrons) that do not generate electromagnetic interactions (i.e. do not emit or absorb light).

1.3. The Universal cosmological model versus the standard model

Are we challenging the standard model in this book ? Not at all, but we are simply saying that the Universal Cosmological Model, which explains the Universe from before Big-bang to man, has just swallowed it up like all the other models. And this is what we tend to demonstrate in this chapter so that it is not a slogan.

The only subject we are trying to clarify in this point is the genesis of the Universe from before the Big-bang. That is to say forget the mathematical and physical theories that are part of the Standard Cosmology model, to focus on the points concerning the origins of the Universe.

The Big-bang theory, we said, deals with the beginning of the Universe, when it was made up of elementary particles and radiation. It is based on Abbot Lemaitre's idea of the primitive atom. According to Lemaître, galaxies are fragments that were projected by the explosion of this atom, creating the expansion of the Universe. This model, corrected by G. GAMOW, is at the origin of the Big-bang theory on the evolution of the Universe. It constitutes the overall theoretical framework in which all the events of the Universe take place. Thus, it is a model that is strongly enriched by other theories that followed afterwards, known to this day under the name of the Standard Model.

Instead of the primitive atom., Gamow assumes that the Universe was born from a gigantic explosion and that the different elements observed today were generated just after this explosion called the Big-bang. This is what science has finally retained in place of the explosion of the primitive atom. by Abbe Lemaitre.

According to G. Gamow's theory, which is accepted by science, everything should be analyzed in relation to the Big-bang, which is taken as the starting point of the Universe. Hence the introduction of Planck's time and inflation where we should interpret in a very short space, events whose duration for us do not correspond to their linking time, but simply to conform to this theory which has become the standard model. I am not alone in expressing such doubts. Let's read this scientific thought :

*"Similarly we are unable to prove the scientific theories explaining the origin of the planet or the creation of an atom – but we accept, under the weight of evidence, that these theories are correct». (*Cf : Linda Gamlin, The origins of life, ed. Artis-Historia, Brussels. 1989,P. 32).

Planck's time is most often used as an order of magnitude characterizing phenomena still poorly described by current physics. It is generally rounded to 10^{-43} seconds ; as the time we can access. Beyond this limit, the laws of physics cease to be valid. This time extends to the time when the four fundamental forces were separated, at 10^{-35} seconds, bringing the Universe into the time of inflation. Beyond Planck's time, the description of what the Universe is, no longer a material of physics according to current science, in the sense that these phenomena by nature escape observation.

After this scientific limit, we literally entered the realm of metaphysics. For the philosophers who try to think about it, metaphysics is something of the waves which does not have detailed diagrams ; everyone can approach it wherever they want. At present, quantum mechanics is trying to unravel these mysteries.

The Universal Cosmological Model, on the other hand, is the scientific restitution of the biblical creation story. This is to say that we are still in physics and not only in religion, because, Moses described not only all the great sets of what happened before the Big-bang, but the mathematical implications as well. At least, even if today we cannot quantify it in time and space, we have the verses as "the spirit of God moving upon the waters... the light was" (Gen 1:2-4). The spirit of God representing the original quantum energies and the waters representing the original quantum particles. Taken from this angle, no astrophysicist can deny the scientific significance of these verses written 1500 BC.

The enigmatic moment when the Universe began to expand is called "Planck's Wall". As I said, this is because the known laws of current physics stop there. So Planck's wall can then simply be described as the moment at which the current standard cosmological models no longer allow us to know what comes before.

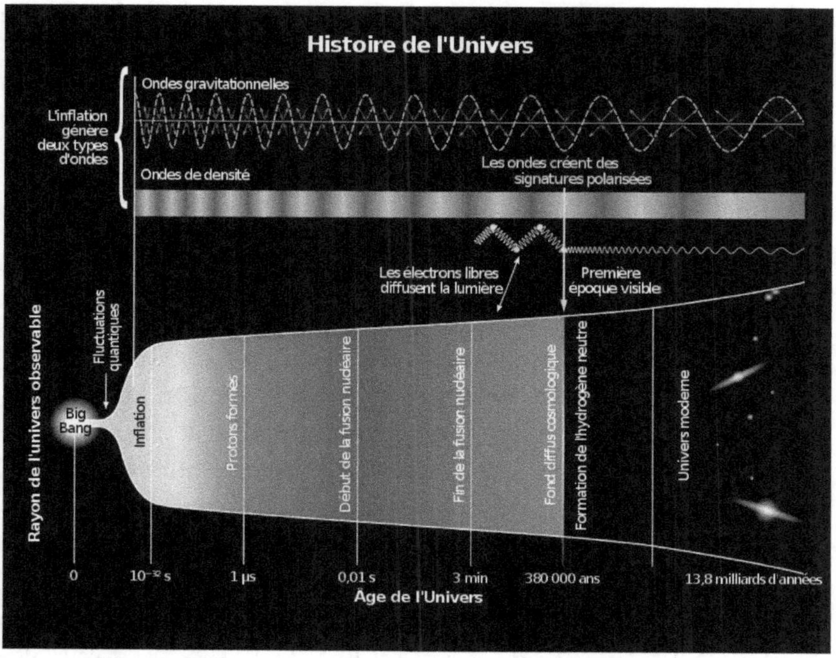

The Planck era is located just after the Big-bang (left) or "the separation of the waters from the waters", but before the era of inflation (in beige). (Source : Wikipedia; the Planck Era).

To understand more about the different Big-bang eras up to the hadronic era (between 10^{-6} and 1 second after the Big-bang), i.e. the era of the formation of the nucleus of the atom, is something that gives me headaches that I did not want to pass on to my readers. All we can conclude is that after the Big-bang according to science, the Universe experienced in the first fraction of a second of its existence an extremely rapid and violent expansion, a step called cosmic inflation.

This theory was first address in 1979 by the American physicist Alan Guth and confirmed according to some articles by the BICEPS2 telescope in May 2014. (Cfr : www.tdg.ch/preuve-inflation-cosmique-big-bang-faite).

But, one year later, this confirmation was rejected by the CNRS on the basis of the results collected by the European satellite Planck with this press release :

"Too bad for the astrophysicists who track down the slightest index to confirm this step of inflation... Since the announcement of the Planck satellite results, the status of inflation has thus become theoretical again». (Cf : sciencesetavenir.fr/lapreuve-de-l-inflation-de-l-Univers-n-a-toujours-pas-ete-observee ; by Azar Khalatbari, 02.02.2015).

For a good understanding of Planck's time and inflation by my readers ; it is an explanation that serves to justify, for example, how physiologically a being is 120 years old, whereas according to logic he was born only five years ago. So we create a formula to explain its growth in this time span so that we don't get within the 5 years that marked the time of creation. But, at zero seconds, we find that this being was already born. We are now faced with two hypotheses ; either we enter into the sub-multiples of seconds, or the being was born before the five years since the creation of the Universe.

In the case of the Big-bang, in order to explain the creation of the elements, in the standard model, we adopted the first hypothesis by entering the sub-multiples of seconds and choosing 10-46 seconds as the zero instant of the Universe or the Big-bang.

As I said, many scientists accept this only by convention, as Avi Loeb, a Harvard-Smithsonian physicist, said : "The current situation with inflation is such a flexible idea that it cannot be experimentally falsified". The latter mentioned whether or not cosmic inflation has taken place (which is also a material of debate in the scientific community), but this is also true with regard to the timing of the event, concluded the author of the article. (Cf : trustmyscience.com/new-etude-matiere-noire-plus-ancienne-que -big-bang,Stéphanie Schmidt, 9 August 2019).

For the Bible, on the other hand, through its scientific restitution in the Universal Cosmological Model, the Big-bang does not mark the beginning of creation and this took place on the second day characterized by the phrase "the separation of the waters from the waters". Let's read on :

Ge 1 : 6And God said, Let there be a firmament between the waters, and let it separate the waters from the waters. 7And God made the firmament, and separated the waters that were under the firmament from the waters that were above the firmament. And it was so. 8 And God called the expanse heaven. And there was evening and there was morning : and it was the second day.

So for our previous example, taking biblical time into account, the one hundred and twenty-year-old will have, despite his exponential growth, some of those years before he is born on the first day. We will not have to create a limited Planck's time for him to justify his age, since a biblical day is an evolutionary period that has a beginning and an end. So, between science and the Bible, which one is the logical choice for this example ?

Scientists tell us that before the creation of the luminaries (Ge 1:14), things happened at speeds that exceeded the speed of light during the inflationary period, which means that these things happened simultaneously with the Word. So why reject sentence such as "God said and the thing was» ? By accepting the theory of inflation, scholars confirm this biblical saying which covers the events that took place before the Big-bang where there were no mathematical notions of space and time, and the Big-bang (separation of the waters from the waters) which took place in a spontaneous way.

The Italian physicist Gabriele Veneziano, the creator of string theory, also questioned, not Planck's time, but the limit that was given to it. In an interview with Libre Belgique, he said the following :

"The traditional cosmology of the Big-bang comes up against a serious knot. If we go back to the past, we notice that the Universe in Planck's time was considerably larger than it was supposed to be. The overall size of the Universe at that time was of the order of a millimeter, 30 orders of magnitude too big (it was 1000000000000000000000000000000 times too big) compared

to the length of Planck (which is the distance travelled by light in Planck's time)...The Universe was therefore too big for its age. An attempt was made to solve this with the theory of inflation imagined by Alan Guth and Andrei Linde, also present at the Solvay Council. They introduced the idea of an ultra-rapid expansion of the Universe, just after the Big-bang... To resolve this paradox, we must therefore either admit inflation or imagine that the Universe is actually older than we think and this raises the question of the beginning of time. » (Cf : La Libre Belgique 2005. Guy Duplat).

As we can see, when it comes to interpreting one event to conform it to another, even science enters into the illogical field. Now, Abbe Lemaitre's idea of a primitive atom. already assumed a pre-Big-bang, because this first one is itself the combination of other original particles. The discovery of quarks, if Lemaitre's theory had been developed, would have allowed science to go beyond the Big-bang. But the twentieth century was a communist century in which material ideas and the rejection of a creative God prevailed. And, everything was developed taking into account the theory of the American scientist of Russian Origin G. Gamow, who replaced the notion of the primitive atom. with that of the Big-bang.

Therefore, the Universal Cosmological Model is a potential model in cosmology and not a historical model, because not everything has been discovered yet. The before Big-bang which is described in the first day of the biblical creation and scientifically restored in the Universal Cosmological Model is not yet attested and validated by science. But it is only a question of time. A re-reading of the different results of the diffuse cosmological background is necessary. This is the approach undertaken by the theoretical physicists Roger Penrose and Vahe Gurzadyan, whose results I have developed in the section concerning the Conformal Cyclic Cosmological Model.

Indeed, when we analyze a result with the idea that we must take into account the time of the Big Bang, we already create a limitation in its interpretation.

The Universal Cosmological Model does not deny what science says about the post-Big-bang period, i.e. due to the expansion of the Universe, the extremely high temperature and density decreased to below a certain limit, and this allowed the subatomic particles to fuse and begin the creation of the simplest chemical elements, such as hydrogen, helium, deuterium or lithium. The gradual cooling of the Universe due to expansion, leading to the creation of the light elements, forms the basis of Gamow's theory. Today, it is according to the Big-bang theory that the constituent elements of the Universe are specified, as well as their transformations from the initial explosion.

So, according to this theory, an initial explosion that occurred some 15 billion years ago in a then extremely dense and hot Universe is at the origin of the expansion and structure of the Universe as we see it today. The Big-bang theory is therefore an evolutionary model which is opposed to static and stationary cosmological models and to which the Universal Cosmological Model also subscribes. It is this Lemaitre, Friedmann and Gamow trio theory which has become the standard model in cosmology and accepted by the majority of cosmologists.

All the other unanswered questions in the standard model should be answered if science looks at the Universal Cosmological Model. Thus, the Universal Cosmological Model, whose a part of each model is inside, is not an extension of the scientific Standard Model. It is an old model, because it is contained in the Bible, but it has become the new model following its scientific restitution called "the Universal Cosmological Model".

1.4 The perspective of the Universal Cosmological Model on Dark Energy and Dark Matter

The biblical scheme of creation also says nothing about the origin of dark energy as science does. It presents it as the state of the Universe at its beginning, i.e. before the Big-bang and not after the Big-bang as science invalidates it. Our argument is drawn from the verse : "The earth was formless and empty : there was darkness on the surface of the abyss" (Gen 1:2). The darkness here has nothing to do with night, but rather was energy at the very beginning of the Universe. So, if its reaction as we will see goes back to the Big-bang, it was already part of the very state of the Universe before the Big-bang.

As for dark material, we have referred to what happened when the first biblical light emerged before the Big-bang and what followed afterwards as we can read :

Gen 1:3 God said : Let there be light! And there was light. 4 And God saw that the light was good; and God separated the light from the darkness.

God being invisible as a hero in the shadows, this name covers all the physical processes that cannot be described at the time, but took place according to his will. Light and darkness being immaterial, if they were separated, there was an immaterial barrier also to separate him. In science we speak of the interaction between light and darkness. This is how we suspected the presence of the dark matter which separated them and which would also be a material from before Big-bang. This material would have appeared just before the Big-bang and the emergence of the first light at the moment when a spirit of God (an unknown energy : UPF) was moving the waters (Ge 1:3). The waters in this frame represent the original particles.

It was a great surprise for me to see that science is beginning to follow in my footsteps without some scientists having read my book. What is the consequence if dark matter had really appeared before the Big-bang? The answer comes from the astronomer and physicist Tommi Tenkanen of Johns Hopkins University who said :

"If dark matter is composed of particles formed before the Big Bang, they affect the way galaxies are distributed in the sky in a unique way. This connection can also serve to reveal its identity and bring new insights into what preceded the Big Bang... if dark matter was really a remnant of the Big Bang; in many cases, researchers would have already seen a direct signal from dark matter in various particle physics experiments". (Cf :trustmyscience.com/new-and-old-dark-matter-big-bang/9/08/2019 by Stephanie Schmidt).

So, even if scholars had doubts about believing my arguments based on biblical verses, astrophysicist Tommy Tenkanen has just made the assumptions that allow them to think rationally.

1.5. The theory of the primitive atom. versus the conception of the Universal Cosmological Model

Let's start with the idea of the primitive atom. of the Belgian abbot Georges Lemaitre. For him, this atom exploded because of instability. This effectively demonstrates that he did not make the connection with the history of creation to discover whether his theory is biblical. If he had, he would not have used the scientific word "unstable". Moreover, his contemporaries described his theory as based on faith, which he rejected outright, saying that his theory was purely scientific and had nothing to do with religion. He even rejected the congratulations of Pope Pius XII by making him understand that his theory had nothing to do with religion.

Indeed, in his theory, the Belgian abbot Georges Lemaitre assumed that the Universe had once been condensed into a huge mass, which became unstable and exploded. (Cfr : Damian Grint, les mystères de l'Univers, ed. Christophe Colomb, P. 146, Paris, 1985).

However, one cannot speak of instability if it was part of the results requested by its author. As the final outcome of the Big-bang was the emergence of man, we say that the Universe exploded at the moment it reached its first result willed by God and that this was not due to a chance of nature or instability as G. Lemaitre concluded.

To speak of the instability that caused the Big-bang, G. Lemaitre concluded that it was not due to a coincidence of nature or instability. Lemaitre was surely inspired by the principle of what would happen to the functioning of the atomic bomb. In this principle, when the uranium-235 nucleus is bombarded with neutrons, it becomes unstable and splits into two, three... creating chain reactions, releasing a phenomenal amount of energy, while causing a strong explosion. In these reactions, we are in front of a stable nucleus which is bombarded by a neutron to create instability. This idea can be completed by this quotation :

"It was assumed that the entire mass of the Universe was originally concentrated *in a very small sphere with a temperature of several billion degrees. When this "atomic bomb" exploded, its pieces were ejected with relative velocities sometimes close to that of light... the fifth consequence of this rarely explained model is that the state of the Universe at the "singular point" requires a divine creation...".* (Cf : http://www.matierevolution.fr/spip.php ? article, Big bang or not Big Bang, 15/05/2009 by Faber Sperber, Robert Paris*)*

The Universe has undergone an evolutionary process that ended with the formation of man. Therefore, we cannot speak of instability, as long as the process was transitory and that it ended the second biblical period before the third began. This is how Moses in the Bible used the verb to separate (God separated the waters from the waters) even if there was an explosion or expansion to demonstrate that it was not an accidental phenomenon, nor an instability as Abbot G. Lemaitre thought it was.

The theory of G. Lemaitre was implied, starting from the standard model that begins the creation of the Universe from the Big-bang when this atom began to expand as we can see in the diagram below :

1.5.1. Big-bang scenario

Before 10-43 seconds : Superforce.

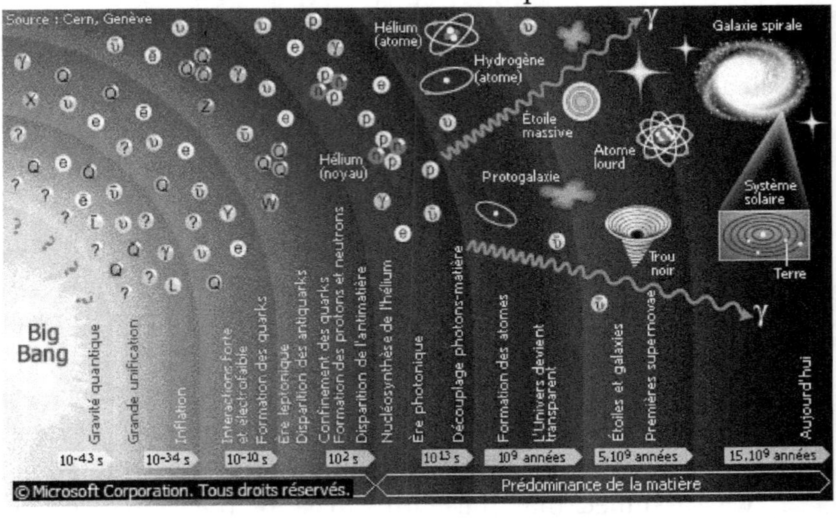

"According to the Big-bang theory, a theory accepted by the majority of cosmologists, the Universe was born about 15 billion

years ago in an unimaginable explosion called the Big-bang. Based on what was there before this explosion and immediately after, exactly 10^{-43} s, current theories, which are numerous and controversial, do not yet make it possible to say. But from that moment (10^{-43} s) to the present day, the Big-bang theory, supported by recent observations, notably that of background radiation from the sky by the COBE satellite, makes it possible to describe the major stages in the evolution of the Universe" (Cfr : Big-bang theory : encarta© Microsoft Corporation. All rights reserved).

In this diagram of the cosmological standard model, we see how science conceives the origin of the Universe from the Big-bang. We have said in the Universal Cosmological Model that the Big-bang is found in the Bible in Genesis 1:6 "the separation of the waters from the waters" and this was only a transitory phenomenon. That is why those who read the Bible can only consider the current standard model as a transitory model. The true pattern can only be found in the Universal Cosmological Model, which reproduces the biblical pattern and which, as I have said, is not an extension of any model but encompasses them all. It is a model which originally comes from the Bible and is not inspired by any scientific model.

While taking into account the theories that have enriched the standard model since Einstein up to these days, we nevertheless ask for a complete revision taking into account the Universal Cosmological Model. The Universal Cosmological Model thus becomes the foundation of all cosmological models.

When something is swallowed, the organism takes what is good to be part of it and expels the debris by forced means. In our case, this is an imposing constraint to some theories to conform to the current cosmological standard model.

In the scientific picture above, we can see that after the Big-bang, there are only question marks. Science does not want us to talk about God, but they want us to accept the word riddle or question mark. These words for us believers also represent God, although today in the debate about the Universe, there is no longer room for thought to publish a theory. There are now laboratories and very powerful instruments where every theory is scrutinized.

1.5.2. Scenario of the Universal Cosmological Model

In the Universal Cosmological Model, when speaking of the events that took place before the Big Bang, if we stick only to the physical manifestation without touching where faith intervenes ; the Jerusalem version of the Bible says in Ge 1:2 "... a wind of God moved over the waters".

Anyone, including astrophysicists, will say that a "wind of God" is an energy whose author attributes its origin to God, but scientifically it is energy. Quantum physics also confirms this by talking about the energies of the vacuum before the Big Bang. It is like when we attribute creation to God, whether we agree or not, creation is there.

So, here we already have energy and original material called "water" by the Bible and Universe by science. And, as a result of the movement of this energy on this original material based on this same verse, there was first of all the emanation of light or rays (Gen 1:3). Here we see that we were already dealing, according to science, which does not recognize this period, with a fundamental force qualified as God's wind around this material, the first result of which was the emergence of rays. And this radiation, according to the Bible, could not dissipate the darkness, i.e. the energy that was already present (God separated the light from the darkness). I have already spoken about this on several occasions. This closed the minus one era of the Universe called the first biblical day.

After that it was the Big-bang that came when the temperature was extremely high and the density reached an extreme level. This reflection did not come from ourselves and is by no means a copy of the Standard Scientific Model. It is rather the emergence of light before the Big-bang that confirms it (God says that light is and light was). Light only appears in material if it is heated to the extreme. This was followed biblically by the verse "...God separated the waters from the waters" (Gen 1:6). This phenomenon, which we call the zero instant of the Universe, biblically occupied a whole era called the second day, as we can read :

Ge 1:8 God called the expanse heaven (empty space created by the expansion of the Universe). So there was evening and there was morning; the second day.

As we can see, after the publication of this book, no astrophysicist, not even an atheist, will be able to stand before an audience and declare that the biblical creation story is a myth or an aporia.

We have already explained that the name God intervenes in the physical phenomena of Genesis to demonstrate that this was part of his plan. When you prepare food, you don't go into the pot yourself, but you watch everything from the outside and knead at the right moment. No cook can tell his boss that it was the fire that burned the food. The responsibility lies with the cook. It is the same thing in the Genesis diagram based on creation where Moses recognizes in God the only person responsible for physical phenomena by speaking of "God said", "God made" or "God created".

It is well known that believers attribute everything to God. Even Adam did not spare God by saying in Ge 3:12, "it is the woman you put with me who has just transformed the ideal of the kingdom of heaven into hell". But on the physical level, things had happened naturally according to their wishes. Christians who criticize Adam's behavior are no better themselves blaming God for Jesus' death, attributing it to his will., however, physically Jesus was crucified according to Jewish law for proclaiming himself the Son of God (Jn 19:7). By this we understand that phrases such as "God says" or the "will of God" are inner phenomena, but whose fulfillment follows physical processes.

It is because of these physical processes that the famous astrophysicist Stephen Hawking said that the Universe does not need God to set physical phenomena in motion and that all results are physically being proven. (Cfr : Conference-Documentary "What do you think of Stephen Hawking's statement" by Hubert Reeves).

It is also like nowaday's Jewish leaders who will say that God's permission was not needed for Satan to enter Judas (Luke 22:3) for the crucifixion of Jesus, because it came as a result of these statements. These two examples clarify the concept of "God says" or the "will of God", which are inner concepts that science does not like to be brandished. Fortunately, the Universal Cosmological Model has just clarified things.

Let's go back to the earth. We see that darkness and light and the separation of the biblical waters were not phenomena due to miracles. The pattern was scientifically respected, Moses spoke of the waters, the darkness, the spirit (of the wind) following the teaching language of his time. And, we cannot put a sheet over it as long as it is true. It is not the change of colors or languages that prevailed in his time. It is not also the black phenomena that makes us the first inhabitants of humanity, but it is the analysis of the background.

We know that science has not yet reached the level where it can assimilate the birth of the Universe through a particle accelerator. But, at least for the time being, it has very powerful devices at its disposal to turn back in time and see how events may have evolved. So it's time to tackle the enigma that has introduced many of the falsehoods into the Standard Scientific Model. This is the consequence of what happens when we dismiss God, even though science.

Some astrophysicists speak of an explosion of energy during the Big-bang, but in the scientific documentary "New Unified Field Theory (Explanation of Gravitation)", it is said that energy only transmits information and that heat can only germinate if material is formed. It is up to astrophysicists to decide this for themselves.

As for the material that exploded during the Big-bang, Father Georges Lemaitre spoke of the primitive atom, but his idea was swept away by Gamow who spoke of dense material. Science and the standard model offer us nothing and talk about enigma or energy or condensed Universe. The Universal Cosmological Model sticks to the Bible, i.e. there was indeed an original material called "water" by Moses. Personally, my scales are leaning towards Georges Lemaitre while waiting for science to answer our question.

Am I the only one to question the standard model ? Let's read these scientific writings :

"The Standard Model has proved to be consistent with almost all the experimental tests carried out so far, however, most

particle physicists believe that it is an incomplete description of Nature and that a more fundamental theory is waiting to be discovered. In particular, a model incorporating both microscopic scale gravitation and the standard model without contradiction is still awaited". (Cf : Wikipedia : Subatomic particles).

For us, to talk about gravitation on a microscopic scale is also to see beyond the Planck's Wall, that is to say the process that has brought the Universe back to being able to condense. For science, it is said that the Big-bang marks the beginning of the dilation and expansion of the Universe and that it was not an explosion.

How can dilation or expansion create material if there were no elements (biblical waters) ? In fact, dilation and expansion only play on the outside, while the inside filled up with energies. If the Big-bang was followed only by an expansion of the Universe, there would have been no creation of the elements, and the same original masses would be the same today but moving away from each other. Since this is not the case, then, we fall back to the biblical phrase that says "from energy (spirit of God) over material (waters), there is a rise in temperature and the production of radiation (light). And from this condensed material (waters) there was Big-bang (separation of the waters from the waters). (Ge 1 : 2-7 ; scientifically restored) (Cfr : "Quark." Microsoft® Encarta® 2009 [DVD]. Microsoft Corporation, 2008).

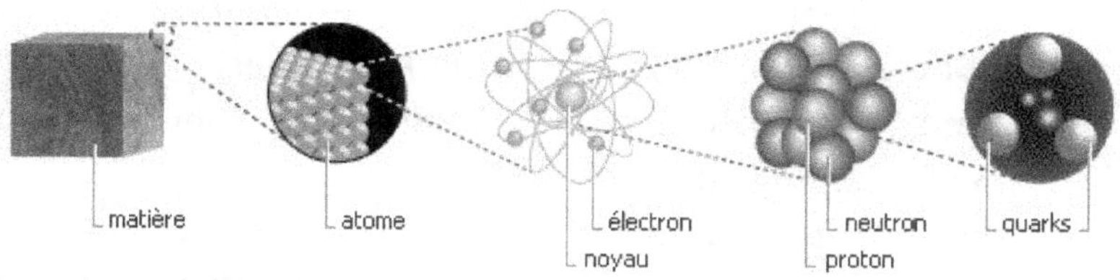

In this diagram, quarks group together in triplets to form neutrons and protons, which in turn group together to form nuclei ; the latter combine with electrons to form atoms.

If the original material contained in the Pre-Big-bang bubble, has not reached one of the components of the above elementary particles, which is what happened before the quarks, there are still other sub-components of material. According to science, we speak of a plasma soup that preceded the quarks. There are no soups without elements. So, if quarks are the last components and they were formed after the Big-bang, it is because the very notion of the Big-bang does not become enigmatic, but imaginary. However, since it is not imaginary, it is up to science to solve this question. Others say that it was an energetic condensation.

It is the search for the answer to this question that will modify the Standard Model and the goal to be reached will become the discovery of the phenomena and elements cited in the Bible. As for the constituent of these elements, as we have said, the study will have to be made in the light that occurred before Big-bang according to the Universal Cosmological Model. And, for the time being, we believe that it is fossil radiation.

In the light of all that is written here about the fragility of the Standard Model theories, we can conclude with these words from Professor Aurelian Barrau :

"The current standard model is indeed unsatisfactory. The predictive capacity is good but the foundations are fragile... It contains many free parameters whose value is completely arbitrary. It is based on many artificial assumptions. It unifies only two of the four interactions. And, above all, it does not include gravitation! Therefore, it is obviously necessary to go "beyond" this model". (Cfr : Les grands Entretiens d'Actu-Philosophia, Interview with Aurélien Barrau, 5 June 2012, by Florian Forestier).

He is not the only one, it is also the concern of all astrophysicists as we can read :

"But the real concern comes from the fact that the standard model is only considered by the majority of physicists as a low energy model which must, from a certain scale, give way to another model which includes more phenomena. » (Cf : Wikipedia : supersymmetry).

This can also be confirmed by this other sentence :

"However, in the current state of knowledge, gravitation is still not included in a quantum field theory. Such a theory would be necessary to go back in time beyond Planck's time". (Cfr : THÉORIE

DU BIG-BANG : encarta).

The only model that includes more phenomena is the biblical creation scheme, first restored literally in the Universal Cosmological Model in order to allow scientists to break down Planck's wall, to penetrate it and tell us what they will find.

Section 2
Cosmic inflation

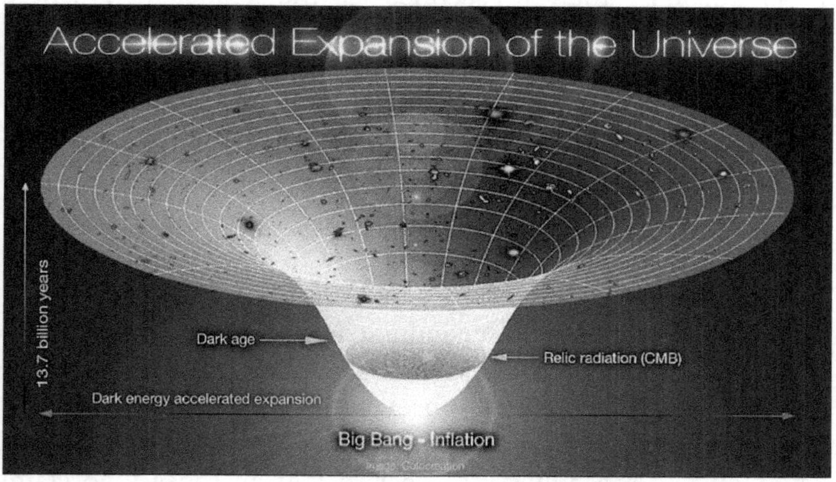

https://upload.wikimedia.org/wikipedia/commons/thumb/c/c2/ Lambda-Big_Bang- Inflation.jpg/800px-

2.1 Introduction to the cosmic inflation theory

Cosmic inflation is a cosmological Big-bang model in which a region of the Universe comprising the observable Universe underwent a very violent expansion step after the primordial explosion, only to continue to increase thereafter at a much lower rate. This step of expansion would have occurred very early in the history of the Universe, at the end of the Planck era, or relatively soon afterwards (at the order of 10^{-35} seconds).

According to the Big-bang model, the expansion step of the Universe was extremely rapid in the first microseconds. This phenomenon has been named cosmic inflation. The theory of inflation was proposed by the American physicist Alan Guth in the 1980s, and developed shortly afterwards by Paul Steinhardt, Andrei Linde and Andres Albrecht.

According to scientists, this theory, which has not yet been confirmed by observation devices, nevertheless solves various cosmological problems, including that of homogeneity, isotropy, flatness (the Universe is flat) and the inclination of the Universe (Euclidean).

The Inflation started 10^{-43} seconds after the Big-bang. Between 10-46 seconds and 10^{-43}seconds is the Planck era where everything is cosmic soup. Planck's time could therefore refer to the period of the Universe which would extend to 10^{-43}seconds from the hypothetical moment zero. Since no theory describes the Universe below Planck's time, the Big-bang, strictly speaking, remains inaccessible to scientists. Thus, the first origin of the Universe remains an indecipherable enigma. Moreover, the Big-bang theory does not describe the events immediately preceding the primordial explosion.

Therefore, within 10^{-43}seconds, it is a dark zone where the current laws of physics no longer answer. This is the divine realm and that science is trying to unravel the mystery with quantum mechanics. Here too, the American Nobel Prize winner Richard Feynman had the honesty to declare that all those who claim to understand something about quantum theory – including physicists – are liars. It's all probabilistic. (Cfr : Blogs zevenzeur ; Introduction to fundamental physics, published on 18/02/2013).

By this we understand that even in the quantum field, for today's science, everything is only probability. As for theology, understanding this concept is part of the field of madness.

2.1.1. Experimenting with Cosmic Inflation

As I said, at present, arguments can no longer be theoretical. Man in his ingenuity as lord of creation manages to create devices that allow him to test these predictions. However, testing certain cosmological theories is often difficult because it requires very high energies that our current laboratories are unable to produce. One way for scientists to solve this problem is to replicate cosmological phenomena using physical systems that behave in the same way.

Indeed, an American team of physicists has been trying to recreate, we have read, the phenomenon of the inflation dynamics of the early Universe in the laboratory. They succeeded in recreating it using a Bose-Einstein condensate. Their results were published in the journal Physical Review X.

This experiment, the journal says, revealed many similarities with the inflation of the early universe, however, differences remain. In the inflation model, the residual energy from the decay of inflation leads to the creation of fermions and bosons, which was not the case during the experiment. In addition, quantum effects became dominant in the EBC (Bose-Einstein condensate), causing it to rotate and change its atomic properties. Physicists understand what this means. (Cfr : trustmyscience.com/cosmic inflation recreated in the laboratory thanks to a Bose-Einstein condensate, Thomas Boisson, 2/05/2018).

What is important for us, is that we do not refute the phenomenon of cosmic inflation, but of the times that we allocated for the creation of the elements during the inflationary period.

Within the framework of the Big-bang model, an idea has recently been introduced as an alternative to cosmic inflation to describe the very first moments of the history of the Universe, the ekpyrotic model. (Cf : String theory : Wikipedia)

2.2. The Universal Cosmological Model versus the theory of cosmic inflation

The theory of inflation confirms in many respects the Universal Cosmological Model, i.e. that all material was originally gathered together in a bubble according to physics and described as "water" according to the Bible. Then the Universe underwent a rapid and brutal expansion which was given the name inflation. This theory does not necessarily confirm the Big-bang as we have said, but the expansion of the original Bubble until it reached a stage known as step transition.

This theory of an original bubble at the beginning, where all material was united, is confirmed by the temperature of the cosmological diffuse background, which is strangely uniformed throughout space, whereas the galaxies are very far apart from each other. In fact, it is this temperature that confirms that all material had originally gathered and confined itself, according to scientists. However, science was also looking for the remnants of this expansion called gravitational waves to confirm the Big-bang. These waves have just been detected by theBICEP2 telescope. For us it is a biblical confirmation of the verse "God separated the waters from the waters" called "Big-bang" by science.

Therefore, the confirmation of the Big-bang is no longer debatable because it is attested by religion, ancient models, and science through the standard model. What remains now is the confirmation of the light that appeared just before the "separation of the waters by the waters" and which closed the pre-Big-bang period.

Once this light has been confirmed, which for us would be the fossil radiation, we will try to answer the question : why, at the appearance of this light, it is written "God separated the light from the darkness» ? Scientifically we can say, what separated this radiation from an already existing energy ? This will certainly lead us to know the origin of dark energy and dark matter.

As we can see, the Universal Cosmological Model is religion only because we allocated it, like Moses, the source of everything in God, but it is a scientific model that needs all the contributions to come up with a unique scheme concerning the origin of the Universe, from the Pre-Big-bang to the living beings.

Am I for the inflationary theory ? Theoretically yes. Indeed, following the scientific documentary "beyond the Big-bang", I concluded that Alan Guth would have been the first scientist to be able to defend what I am defending today.

In fact, the scientists found that the Big-bang theory could not explain the riddle of temperature uniformity in space. The time of Big-bang until today is very close to equilibrate the temperature. The calculations will bring the time back to Pre-Big-bang, i.e. biblical time.

Thus, around the 1980s, the scientist Alan Guth would have been the first to reveal that something really happened before the Big-bang, i.e. the emergence of light caused by an intense temperature.

According to the documentary ”Beyond the Big-bang”, the scientist Alan Guth supported the idea that the Universe came from such a small volume, that within this very small volume there was enough time for the extremes to communicate and for the temperature to be uniformed.

For us, knowing that based on the speed of light, it would bring the calculation for the Pre-Big-bang, he introduced the theory of inflation to conform to the standard model, i.e. where everything should start after the Big-bang. The argument was that, at that time, one could not yet talk about the speed of light, because after the Big-bang, events took place faster than this speed. And it was called the cosmic inflation that reigned until the first lights appeared.

For us the events that took place during Planck's time are to be reviewed. For example, we are told that : ”starting 10^{-43}seconds after the Big Bang in the inflationary period, the 80 percent of atoms contained in the Universe were formed between the first second and 3 minutes after it”. All this is to justify the fact that everything was created after the Big Bang, and whose hypothesis the Universal Cosmological Model has just put back on the table for study.

Indeed, the subdivision of time during inflation is so theoretical that many people don't really assimilate it to reality. The period of time in which the creation of the atoms took place is indeed what can be assimilated to the inflationary time of 10^{-35} seconds to 3 minutes in which, according to the author of the theory, the 80% of the hydrogen contained in the Universe was formed, boosted by the drop in temperature due to its exponential expansion.

There is another hiccup in the above paragraph. According to the observations, the elements only can be formed when the processes of gravitation begin. This could only be possible in the Pre-Big-bang Universe which was in deflation with exponential confinement, characterized by the increase in temperature as it happens during star creation. This was described by the biblical verse ”a wind of God moved over the water” according to the Jerusalem version (Gen 1:2). By wind of God, one must expect a whirlwind which Moses allocates to God and the phenomenon effectively responds to the functioning of the Universe.

As we can see, while cosmic inflation is logical, the subdivision of time in this theory is not logical. Rather, it a response to the definition of "God said and the thing was" divided into time. It's a good thing that Alan Guth is not recognized as a religious person, otherwise no one would have any faith in his theory.

2.2.1. Evidence refuting the calculation of the elements creation times during inflation

2.2.1.1. The cosmological diffuse background

While the cosmological diffuse background pleads in favor of cosmic inflation, it does not confirm the subdivision of time until the creation of the atom.

In fact, if we go back to the exact idea, we will discover that the cosmological diffuse background has an origin before the Big-bang and would be the biblical "light" evoked by Moses before the Big-bang or the separation of the waters by the waters and this by the sentence "*God said that light be and there was light*". And God called this light Day to demonstrate that this radiance will never fade away again.

Another argument that confirms our conclusion that fossil radiation is a light before the Big Bang reported by the Bible in Ge 1:3 and that also confirms my judgment on the inflationary period is the research done by three astrophysicists, part of whose article published on the Futura-Sciences website we faithfully report :

There are a very large number of models capable of producing an inflationary period of the Universe and the study of fossil radiation is a good way of constraining the theories. Marc Kamionkowski, Adrienne Erickcek and Sean Carroll have just published an article on this subject. Observations from the WMap telescope indicate a rather curious fact which, without refuting inflation, is a bit awkward for this theory. In fact, relatively little is generally said about these measurements revealing that, on average, one half of the celestial sphere has greater thermal fluctuations than the other hemisphere.

The authors of the study tried to account for this fact by imagining that at the very beginning of inflation, the scalar field – inflation – showed greater differences in values than was imagined in the regions of space that are now visible and that have been dilated by its expansion. But unfortunately this does not work! Marc Kamionkowski and Adrienne Erickcek, however, do not hesitate to evoke another consequence of this strange asymmetry in the diffuse background radiation, perhaps related to curvature. It could be a window on what happened not only during inflation but even... before the Big-bang". (Cfr : futurascience/The bizarre asymmetry of fossil radiation : an enigma solved? 02/01/2009 by Laurent Sacco).

So, by studying the disparity they found in this radiation according to the above article, they came to the conclusion that for a logical explanation, the situation should have started before the Big-bang. Such a discovery cannot be the subject of current interest among scholars until the publication of this book. And, it is now that all scholars will run to search for their work.

Section 3
Multiverse theories

3.1. String theory

Bringing out an explanation of string theory in a literary work is not easy, as the views of the scholars themselves are shared. At present, in fact there are several string theories that have in common a 10-dimensional Universe (nine of space and one of time), or even 11 dimensions. However, we can offer a synthesis for the readers.

In this theory, fundamental particles are considered not as hard balls but as waves on a string. So far, astronomers agree that in our Universe, there are 4 dimensions, 3 of which are space and 1 of which is time. For string theorists, there are other dimensions that we don't see and that vibrate in other Universes.

3.1.1. String theory and the multiverse

In fact, outside our Universe, scholars think that there are still other Universes; which casts doubt on the biblical scheme of a single Universe created by God. And, if we do not talk about it, the conception of a unique Universe will be bequeathed to antiquity. Those who address such arguments are not simple philosophers, but the majority of astrophysicists, great intellectual luminaries ready to demonstrate this with theories.

Some scientists even say that we can be connected with these Universes through the exit of a black hole whose bottom we never see called wormhole and others support the existence of parallel Universes. They even go so far as to say that in each Universe we have our correspondents, i.e. by being here, there is still another self in another Universe. (Cf : Wikipedia, Multivers).

Multiverse means multiple Universes or parallel to our Universe. The existence of these multiverses, is better explained by the string theory. In this model, the birth of the Universe and its initial expansion are due to the collision (by some kind) of a brane and an antibrane, which releases the energy necessary for the expansion of the Universe. This model predicts the possibility of further collisions, which would lead to the creation of other Universes.

As for the explanation of these two names, let's read the following lines in string theory :

"A brane, or p-brane, is an extended, dynamic object, possessing energy in the form of tension on its Universe volume, which is a charged source for certain interactions in the same way that a charged particle, such as the electron for example, it is a source for electromagnetic interaction". (Cf : Wikipedia : Brane).

To enlighten my readers in everyday language; by extended object, we can take the case of a guitar string which, by playing it in a continuous way, emits vibrations that follow in space, creating a harmonic rhythm of the music that follows one another. In the case of string theory, which scientists already begin by saying that it is the best theory to explain the origin of the Universe, this frenzied rhythm creates other Universes that follow as in the frequencies in the diagram below.

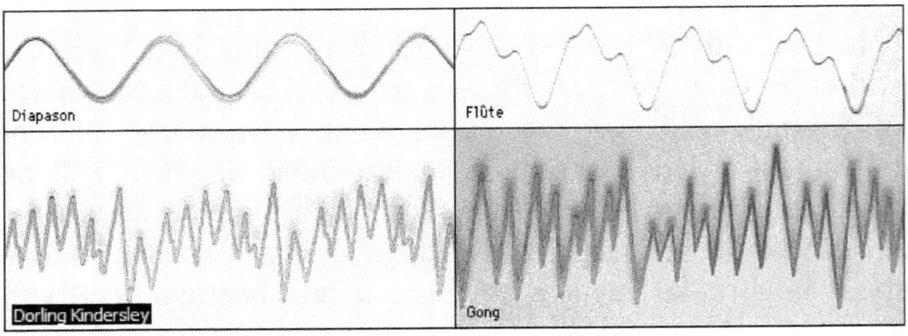

Sound waves from musical instruments

The waves corresponding to the violin, flute and tuning fork have the same peak-to-peak distance when these three instruments play the same note. Dorling Kindersley. Microsoft ® Encarta ® 2009. 1993-2008 Microsoft Corporation. All rights reserved.

This notion of waves comes from Albert Einstein who spoke of gravitational waves. The existence of gravitational waves is predicted by the theory of general relativity : masses moving in relation to each other could lose part of their energy, which would be emitted into space in the form of waves with a speed close to that of light. (Cfr : gravitational, waves." Microsoft® Encarta® 2009 [DVD]. Microsoft Corporation, 2008).

For the proponents of multiverse or multiple Universes, by the collision of a sort of brane and antibrane, Universes are created and follow each other, without one encroaching on the other as shown in the above diagram of sound waves. These waves, emanating from the cosmic strings, were even evoked as a possible source of the large structures of the Universe.

This philosophy is supported by the proponents of parallel universes, as can be read from its interpretation according to Prof. Everett HUGH :

"Every time a particle produces a wave function; this particle generates a parallel world for every possible path. This hypothesis would presuppose that at every moment a multitude of "worlds" come into existence, and that these worlds proliferate in an absolutely incalculable way". (Cfr : Wikipedia : Multiverse).

Thus, for the physicist Hugh Everett, the "theory of multiple worlds" presented and developed in 1957 by him, supposes that our world coexists with many other universes, which are continuously divided into divergent, different and inaccessible universes. According to Everett, each world contains a unique version of each person (each observer) experiencing a different situation at the same moment in time.

3.2. The self-replicating and self-accelerating Universe theory

Another physicist, Andrei Linde, in developing the formulas of the American physicist Alan Guth, on the subject of cosmic inflation, his research led him to suppose the existence of "eternal inflation", which has neither beginning nor end.

According to this theory, there may be an infinite number of "bubble universes" produced in this way, unconnected to each other, or only marginally so. In this case, there is no reason to think that the laws of physics are the same in every Universe, and there would thus be many more Universes than are conceived by us.

Professor at Stanford University, cosmologist Andrei Linde discovered that the Big-bang theory is problematic. In 1982, he formulated a new theory of the Universe, which is an "improvement" on the Big Bang model.

Linde criticized the Big Bang theory for many problems, the most crucial was that the size of the Universe was too big. This is why he proposed the theory of a self-reproducing inflationary Universe.

According to his vision, instead of cosmic inflation, we should rather speak of chaotic inflation; according to the growth of the Universe at its origin it would have been extremely fast. Whereas the classical Big Bang theory describes a soap-bubble-like Universe gradually inflating, Linde's theory describes a bubble-like Universe that would produce identical bubbles, and so on. Linde's Universe would give birth to new Universes through self-replication.

Still for the Russian physicist Andrei Linde, we should imagine the Universe as a set of interrelated bubbles that develop fractally (each part of the whole looks like the whole). Thus, there would have been the creation of a Universe from several bubbles that were formed independently. These new bubbles would in fact are points in the Universe that would have expanded in themselves, without affecting the original Universe. Each of these Universes would have its own laws of physics and could give birth to other Universes, and so on. This mechanism would give rise to a self-replicating Universe that would be eternal and infinite in time and space.

Still in Linde's model, our Universe would therefore be only a tiny part of a "multiverse" that would have no beginning and no end and would be eternal. For some physicists, this new framework, called "quintessence", is fascinating because it opens up the mathematical and physical possibility for creating ex nihilo, from fluctuations in the quantum vacuum.

3.2.1. The Universal Cosmological Model versus string theory, multiverse theory and self-replicating universes

According to astrophysics the string theory is currently considered to be the best way to explain the birth of the Universe. It seeks to unite quantum mechanics by taking into account quantum gravity and the theory of general relativity. The main particularity of string theory is that its aim is not limited to this reconciliation, but that it claims to succeed in unifying the four known elementary interactions in a universal theory.

String and multiverse theories are also taken seriously by us and we do not quote them for pejorative purposes. They concur with the comments that are made in the Universal Cosmological Model concerning the emergence of pre-Big-bang elements. The same is true in Andrei Linde's theory of self-replicating and self-accelerating Universes.

First of all, this theory of Andrei Linde has just confirmed the biblical verse that says that "in the beginning the earth was formless and empty", and I said that it was a way of defining the Universe by Moses. It was in this emptiness that everything he described for the creation of the Universe took place.

In the Universal Cosmological Model, we talk about the spirit of God or Yang (brane) and darkness or Yin (anti-brane) whose interaction actually produces waves and vibrations. Only, we say that these collisions which cause other vibrations, or even other wave and functions even after the Big-bang, may would have been sources of creation of other Universes if not for what happened when "the Big-bang" took place, disturbing the creation of other Universes, according to the aforementioned theories.

So, a very important event happened, as we have said, which physicists should be looking for and which stopped the emergence of other Universes. It is the Big-bang itself. By pouring the elements that later formed large masses into the Universe, the energy that should form other Universes, would no longer have the power to start other creation processes and would be limited only to the waves. It could also be part of the dark energy that contributes to the acceleration of the Universe.

What we mean is that the Big-bang is rather like a balloon with a hole in it and everything that should generate other Universes according to the multiverse theories, but it is instead poured into our Universe from its axis which remains unknown, causing its eternal expansion. This also explains the non-localization of the origin of many cosmic rays, including waves. And this axis is unknown to us because of the isotropy and homogeneity of the Universe.

By the collision according to the string theory of the defined energy of the spirit of God by the Louis Second Bible and the winds of God by the Jerusalem Bible on the qualified primitive particles of water, another qualified energy or radiation of light was produced. This reaction should be repetitive as it is a natural action that cannot be stopped, hence, it is the idea of eternal inflation and the multiverse which has gained popularity among astrophysicists.

At the end of inflation, that is to say after the Big-bang, the Universe was still extremely dense and hot, and underwent a step of very violent expansion that have allowed it to grow. (Wikipedia : Cosmic Inflation).

In addition, the Big-bang, as we have said, produced large sets that prevent particles and other forces from concentrating at a given point in space as in the origin when the abyss was spoken of. Among these sets we have galaxies and black holes that channel and disperse the generated particles and other forces.

So rather than creating parallel Universes, string theory could better contribute to the theory of the expansion of the Universe because we said from the beginning that the Universe itself is dynamic. So, our acceptance of string theory, cosmic inflation and eternal inflation rejects the multiverse, but rather the maintenance of pre– and post-Big-bang phenomena.

Moreover, the theories about the existence of multiverse could be confirmed when we consider that the Universe began with a Big-bang. This is not true, because in the Universal Cosmological Model, the Big-bang or "the separation of the waters from the waters" was a transitory stage, followed by inflation for the emergence of the other elements. Therefore, the Pre-Big-bang bubble was an element in the Universe and not the beginning of the Universe, even if we use the scientific term that the Universe found itself at a condensed moment before undergoing a sudden expansion.

So, with the multiverse theories, the Universes should overlap, pushing each other away, but the distance will never remain uniformed as given by the dimension that each can acquire. In the long run, there should be a collision or mixing such as we see in a river because not only of their different masses, but also because of their rotation and translation.

For an example, let's take two things of the same mass, or even the same weight for a good understanding by my readers. Put them in a stream with a difference in distance. A few minutes later, you will notice that the distance between them when they were put down will no longer be the same. There will be no deviations or closeness due to the current or the wind. This means if there are parallel Universes, they cannot remain indefinitely separated from each.

Moreover, the idea of multiple Universes must go hand in hand with the idea of closed or finite Universes, each one evolving with its properties. However, this idea is only supported by a minority of scholars and the expansion of the Universe tends to contradict it. The Universe is only finite by its gravitational force, just like the edges of a whirlpool as in the drawing below :

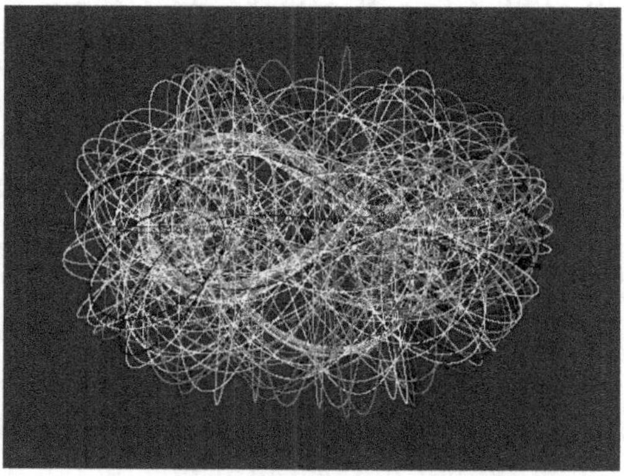

http://supersymetrie.s.u.pic.centerblog.net/e720465c.png

As we have said, for the genesis of the Universe, we are first dealing with masses, and these have generated formulas. Here, I am speaking from the side of humans who are resultant beings. For God, they were first formulas, as the expression "God says" testifies, but creative formulas, and then the masses followed. This answers the formulas of the physics of creation, namely :

– Divinity = In [(Ci .M) + (Po.Ne) + (Ms.Fe)],

– $E=mc^2$

This means that we cannot create the Universe by our formulas. A mathematical formula is created on the basis of observations of constant or repetitive phenomena with a negligible margin of error. However, we cannot assimilate the creation of the Universe because it sprang from the projections of the forces that emanated from God. And if, for the scientists below Planck's Wall, we enter an area where the laws of physics no longer respond to it, in the Universal Cosmological Model, mysticism ended as soon as the name "Earth" was first mentioned in the Bible. In mentioning it, Moses specified that it was "formless and empty". The early universe was waiting for it.

Ge 1:1-2 "In the beginning God created the heavens and the earth. The earth was formless and empty...".

In the string theory, we talk about the dimensions that the universe occupies and that others could generate parallel universes. The proponents of the Multiverse theory, as I have said, have forgotten in their formula that the notion of the axis must be conditional. The Universe by its dynamics and isotropy leaves no place for forming another one and we cannot even glimpse its axis, its equator and poles.

In physics, a system is isotropic if its description is identical in all directions of space. Let's read this scientific reflection :

"What we can already say about singularity is that it is useless to look for it in a precise point of the Universe. The Universe did not grow up around this point, but it is the point itself which has expanded and which today encompasses all space, time and material. So, even if we speak of a point, we must understand that the Big Bang did not happen at a precise point in space, but everywhere at once. Moreover, we can consider that any point in the Universe is fixed, and that it is the rest of the Universe that moves in relation to it. In this case, anyone can consider that it is fixed in relation to the initial singularity, and that the Universe is moving in

relation to it. It is restricted relativity that tells us, in short, that we are all at the center of the Universe". (Cf :astropolis.fr/Le Big-bang : Instant zero or "initial singularity").

The expansion of the Universe is presented today as a balloon which is pumped, the drawing inscribed above becomes more and more distant. These drawings represent galaxies. Thus, the "cosmological principle" postulates that we are not in a particular region of space and that on a sufficiently large scale, the observable cosmos is homogeneous, as shown in the images below taken by the telescopes :

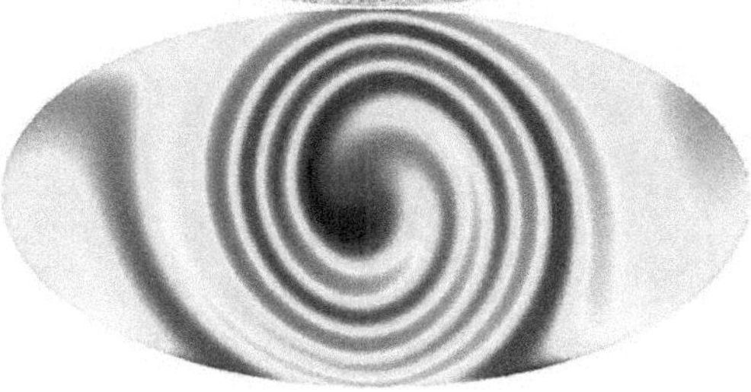

At the top, the map of temperature shows fluctuations in fossil radiation. At the bottom, a simulation shows the influence of a strong rotation of the Universe, which is not observed © ESA, Planck Collaboration, D. Saadeh et al .(Cf : Luigi Bianchi, We would be lost in an isotropic Universe, inwww.futura-sciences.com)

To create an axis and an equator is to go back to a static Universe to which Albert Einstein, who defended it, had already thrown this sheet in the dust bin. Therefore, a parallel or mathematical Universes do not exist.

Section 4
Andrei Sakharov's concept

4.1. Presentation of Andrei Sakharov's concept

In the 1960s, Andrei Sakharov proposed a model of the so-called "twin universes". The first is of course ours and the second is reversed by starting backwards in time (retrochrone) from the famous singularity or "Instant" zero. In this model, the two Universes are disjointed and symmetrical (in mathematics we call them "enantiomorphic") and the times flow backwards in the second one.

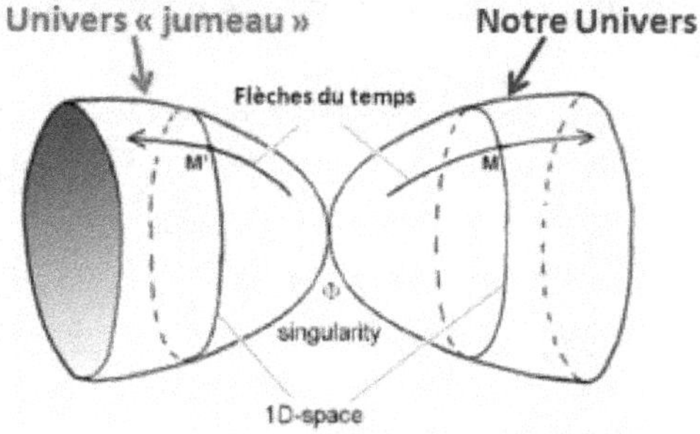

Sakharov's double universe
(https://zevengeur.files.wordpress.com/2018/01/sakharov-mode 03.png ? w=300&h=187)

4.2. The Universal Cosmological Model versus Andrei-Sakharov's concept

We have introduced this section to support an idea that has touched us. Finally, if the Big-bang was preceded by cosmic inflation, it means that the Pre-Big-bang was characterized by cosmic deflation and due to the condensation of water (the original quantum particles) by a wind from God (Ge 1:2). Thus, the laws of relativity worked in the opposite direction.

In celestial mathematics, we have been able to find the figure of **In = 302,752 x 10^{18}km/s and it went into deceleration**.

This thought is similar to those developed by the Russian physicist Andrei Sakharov, except that he spoke of twin universes where time in one was exercised in the other. For the Universal Cosmological Model, the pre-Big-bang deflation supported the Post-Big-bang and expansion.

While I also thought I was the first to speak of the opposite direction of the law of general relativity, I was surprised to discover that the French physicist Jean-Marie Souriau (1922-2012), while trying to solve the problem of Sakharov's double Universe, demonstrated that the inversion of time is equivalent to the inversion of the sign of energy. And, he proposed it as a solution to the inversion of the Twin Universe on our Universe like a sock in the following way :

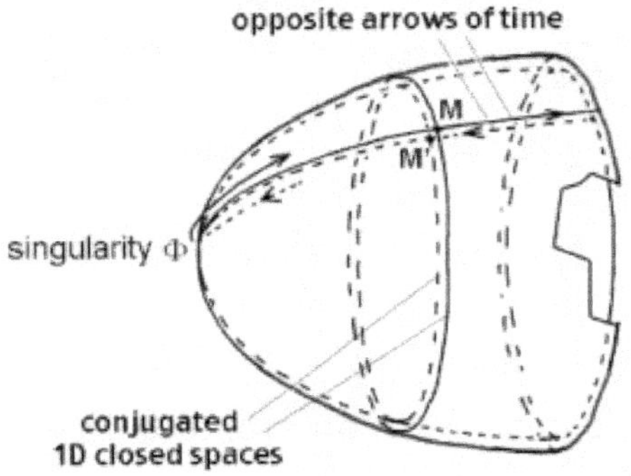

https://zevengeur.files.wordpress.com/2018/01/sakharov-model-janus-01.png

"The two Universes then become one, with a first subset made up of positive energy particles and a second subset made up of negative energy particles, these two subsets interact with each other through gravitation until today. "(Cf : Zevenzeur, www.blogs, zevenzeur, Janus model, the future of cosmology, 12/01/ 2018)

Now in the Universal Cosmological Model, which clarifies all the question marks of the models mentioned or not in this book, we have said that the Big-bang did not only spill material but also laws from before the Big-bang. I spoke of the laws and not of the theory. A theory is a scientific translation of law. Starting from the twin universes of Sakharov and the explanation of the physicist Jean-Marie

We inexorably fall back into the existence of a dynamic before Big-bang or Universal Cosmological Model.

Section 5
The Conformal Cyclic Cosmology Model (CCC)

5.1. Presentation of the Compliant Cyclic Cosmology Model

To understand this Model, let's start by introducing the notion of "cyclical cosmology", or "cyclical universes". Conformal Cyclic Cosmology (CCC) is a cosmological model within the framework of general relativity, advanced by the theoretical physicists Roger Penrose and Vahe Gurzadyan.

The idea of cyclic Universes is based on the theory that our Universe goes through constant cycles including Big-bang and cosmic compressions, unlike the standard model of cosmology, where there would have been only one and the same beginning (Big-bang). In other words, this theory assumes that the end of the Universe is the beginning of a new one. Thus, our Universe would have already been preceded by others and will also give way to another one after a while. The current Universe would be one of the infinite "eons" with a duration of 10^{100}, that is 1 followed by 100 zeros. The word "eon", taken from Penrose, is the translation of the Eastern religious term Kalpa, indicating in these traditions a period of up to 1.28 years Katherine. (Cf : boowiki.info/art/cosmology/cosmology-conform-cyclical).

By combining several Universes in this way, the CCC Model tries to reconcile relativity laws with quantum mechanics. Thus, Penrose assumes that his theory is the only one that can make all the collected data and theoretical formulations such as Hawking radiation coincide, unlike the others, whose theories are considered incomplete, hence the term "conformal".

Penrose, who is a Scholar mathematician and physicist, does not base himself on these thoughts or his philosophy alone. For him, a test of his model would be contained in the cosmological diffuse background radiation, and the streams discovered in it would be the waste products of previous universes, in particular the collisions and explosions of previous supermassive black holes that released huge "bursts of energy".

(Cf :boowiki.info/art/cosmology/cosmology-conform-cyclic.html).

Throughout our book, we have said that the image of the diffuse cosmological background must be re-examined by scientists to discover the traces of the Before Big-bang. Dr. Penrose and his team of researchers have done this and discovered it. Unfortunately, no one was there to teach them that in the first Cosmological Model contained in the Bible, they were already talking about the abyss or black hole before the Big-bang (the separation of the waters from the waters according to the Bible). This is just a parenthesis for all the members of his team who read to me.

So, looking at the cosmic radiation map, Penrose and Gurzadyan noted the presence of a series of concentric circles inside the image (huge black materials), in which the temperature variation is much smaller than expected. The unique aspect of these circles is that, according to the researchers' calculations, some of the wider circles should have been formed before the Big-bang. These are the "Hawking points" that would be visible in fossil radiation, showing the evaporation of supermassive black holes that occurred before the Big Bang, in a world that would have preceded ours.

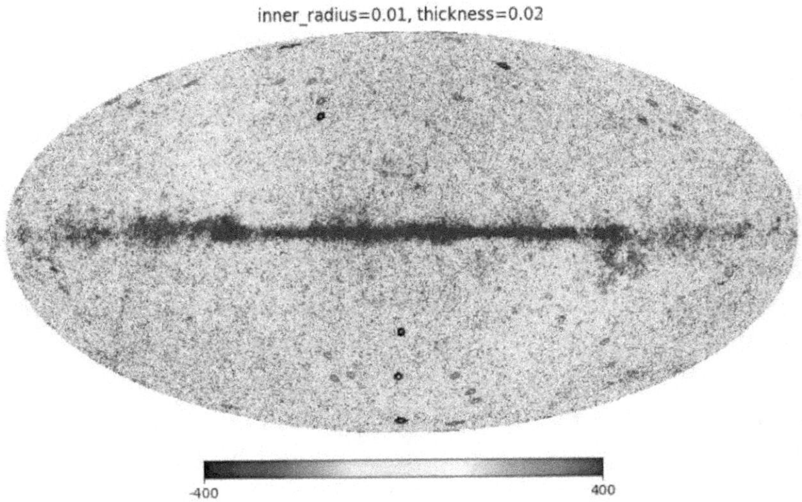

The Hawking points that researchers believe they have found are indicated by circles on this fossil radiation map. Daniel An, Krzysztof A. Meissner and Roger Penrose. (Cf : futura-sciences.com/sciences/actualites/cosmology-proof-before-big-bang-roger-penrose).

This confirms what we keep saying, namely that the image of the diffuse cosmological background still has many secrets to reveal to us, starting with the before Big-bang.

5.2. The Universal Cosmological Model vs. the Conformal Cyclic Cosmology Model

As we can see, Physicist Roger Penrose is working hard to get scientists to accept his Model. He has written a lot and continues to exhibit the evidence of a before Big-bang. We encourage him because the existence of the before Big-bang with dynamic phenomena as described in the Bible as "A wind of God moved over the waters" is part of the Universal Cosmological Model.

How does Roger Penrose's vision of the before Big-bang vision of the Conventional Cyclical Cosmology Model and the other Models differ from the Universal Cosmological Model?

For Roger Penrose, we said, the Universe is eternally expanding – compressing and, paradoxically, with an infinite number of Big-bangs. From where, we don't know where the starting point or the beginning is. It is the same for all the other models that defend the Pre-Big-bang. Almost all of them, according to the vision received in their mind without knowing it and which they have not been able to interpret; there was a collapse (abyss) before the Big-bang.

Not knowing the biblical scheme of creation, they thought if there was a collapse, there would be a black hole. And, if there was a black hole, then there would be a star exploded in a multi-dimensional Universe. And, where did this multi-dimensional Universe emerge? Their conclusions led to the existence of successive universes or multiverse universes.

The Bible, which is available to everyone, offered the most logical answer to their questioning. For the Universal Cosmological Model, scientifically restoring the biblical scheme of creation, the Universe does indeed have a starting point biblically matched by the verse : *"In the beginning God created the heavens and the earth. The earth was formless and empty"* (Gen 1:1-2). It was in this empty space that the first manifestations of the elements took place, as studied in quantum mechanics, whose combination after myriad collusions and mergers resulted in the first energy mass digging into space-time, referred to in the Bible as the "abyss". Seen in the distance, according to the biblical description *"there was darkness on the surface of the abyss, and the Spirit of God was moving over the waters"* (Gen 1:2), we have the image of a black hole.

I had already said that in the dictionary the abyss is defined as a hole where mysterious phenomena take place. This definition does not come from me, I drew it from the glossary of the Free Bible and I myself shouted for joy while reading it because it confirms the existence of a black hole or something similar before the Big-bang as I had restituted it before reading it.

Moreover, concerning the cosmological diffuse background, for the Universal Cosmological Model, this radiation could be the same as what appeared in the Bible at the end of the first period (Ge 1:3), but in a condensed Universe, it will only reappear 380,000 years later after the Big-bang according to observations.

The study must be based not on its reappearance but on how it can appear and subsist until today. Moreover, Roger Penrose's successive Universes could not leave room for several laws of quantum mechanics to transit energies through particles-antiparticles, material-anti material, etc. Until the elements were created.

Apart from these literal reasoning's, this question of successive universes was already debated in the past as we can read in the following quote : "In the 1930s, notable physicists such as Albert Einstein and Richard Tolman, considered the possibility of a cyclic model of the universe as an eternal alternative to the expanding universe model. However, Tolman's work in 1934 showed that these ideas seemed to fail because of the second principle of thermodynamics : it states that entropy can only increase in a closed system. (Cf : https://fr.wikipedia.org/wiki/Mod%C3%A8le_cyclique)

For believers who do not master what the successive Universes mean, in religious terms it means that there were several ends of the world before we arrived at our world. And, after the end of our world, there will still be the appearance of another physical world or other physical worlds which will also experience, after several eons, their end of the world. This is where the thinking of the great scholars leads when they think without religion.

The fields of physics and mathematics being well mastered by Roger Penrose, I only invite him to think on these verses written 1.500 BC and to have all that he wrote for the other Universes transferred to the only Universe before our Big-bang. Whether it is the multiverse, the multidimensional Universes, and the successive Universes, nothing can appear simultaneously with all these structures as in Fred Hoyle's stationary model. It should always have a beginning somewhere in emptiness. And this is what the Bible confirms in the verse "The earth was formless and empty". Something that is formless and empty, we have said, is something that has no material existence except energy. Let us read :

Gen 1:1In the beginning God created the heavens and the earth.2 The earth was formless and void, there was darkness on the face of the deep, and the Spirit (a wind) of God moved over the waters. 3 God said, 'Let there be light. And the light was.4 And God saw that the light was good; and God separated the light from the darkness.5 And God called the light day, and the darkness he called night. So there was evening and there was morning, and that was the first day.

All that is written in bold has been scientifically restored in Volume I. No one could have thought that one day understanding the biblical scheme of creation would become a goal to be achieved by scholars for the understanding of the Universe. I too used to read these verses in a literal and meditative way. I did not know that this scheme hid the greatest scientific secret of the beginning of the Universe.

On the scientific level, it is rather the Italian scientist Gabriele Veneziano, creator in 1968 of the theory of the strings, who without knowing it, joined the Bible with his model of the strings which stipulates that the Universe comes to us from an infinite emptiness (cf : lalibre.be.qu'avait-il avant le big-bang ? le 15/12/2005 by Guy DUPLAT).

Section 6
The Electric Universe

6.1. Presentation of the theory of the Electric Universe

The theory of the electric universe was refined by WALLACE Thornhill and David Talbot. According to this theory, electricity can explain all cosmic phenomena in a much more satisfactory way than the theories we already have. Thus, in our Universe, electricity governs all phenomena. Plasma permeates the entire Universe with its filaments. These filaments, in turn, conduct the electric current, and the electric current directs the universe. Therefore, electromagnetic phenomena, gravity, the movement of atoms are manifestations of electric force on different scales.

The electric current flowing along the plasma filaments forms and animates galaxies. This current passes through the stars, making them shine like light bulbs. Electricity is responsible for the formation of planets. The craters that are sometimes found on their surfaces come from electric arcs that, like lightning, imprint various shapes on the ground.

Moreover, according to the theory of the Electric Universe, black holes do not exist, nor does dark matter and dark energy. The Big-bang never happened and Einstein's theory of general relativity is just a big joke. (Cfr : motherboard.vice.com/a-la-rencontre-des-partisans-de-la-theorie-de-l-Univers-electrique by Sarah Scoles Feb. 22 2016).

The title of this theory reminds me of the first dream I had when God showed me how the elements were transformed from the invisible to the visible and I've already talked about it.

6.2. The Universal Cosmological Model versus the theory of the Electric Universe

The much-criticized theory of the Electric Universe is gaining supporters in every corner of the world. But, by rejecting what is taken for granted, i.e. the existence of the Big-bang recognized by the Universal Cosmological Model, it is wrong.

After this little presentation, I would say that the authors of the theory, criticized by all astrophysicists, are not wrong to think like this. What I wanted to correct here goes beyond what these theories say.

Indeed, electricity derives from the word exchange or displacement of electrons. Without the exchange of particles through their intrinsic properties, the Universe will not be what it is today. This exchange first creates electricity (e.g. electrified by friction).

According to the Principles of Creation, a branch of Rev. Dr. S.M. Moon's book The Divine Principle, the intrinsic properties of particles come from the dual characteristics of God who is the cause. These properties, which tend to unify the projected energies by creating a rotation, explained biblically by the phrase "...a wind of God moved over the waters" (Ge 1:2), creates electrical phenomena, hence the basis of the theory according to the Universal Cosmological Model.

Therefore, even before we could speak of electricity as we know it today, in nature, there were already characteristic electrical phenomena created by the crossing of rays. At the level of the deep Universe, astrophysicists observe the same electrical phenomena which they are unable to explain, in spite of those occurring in the Earth's atmosphere and explained by the following diagram :

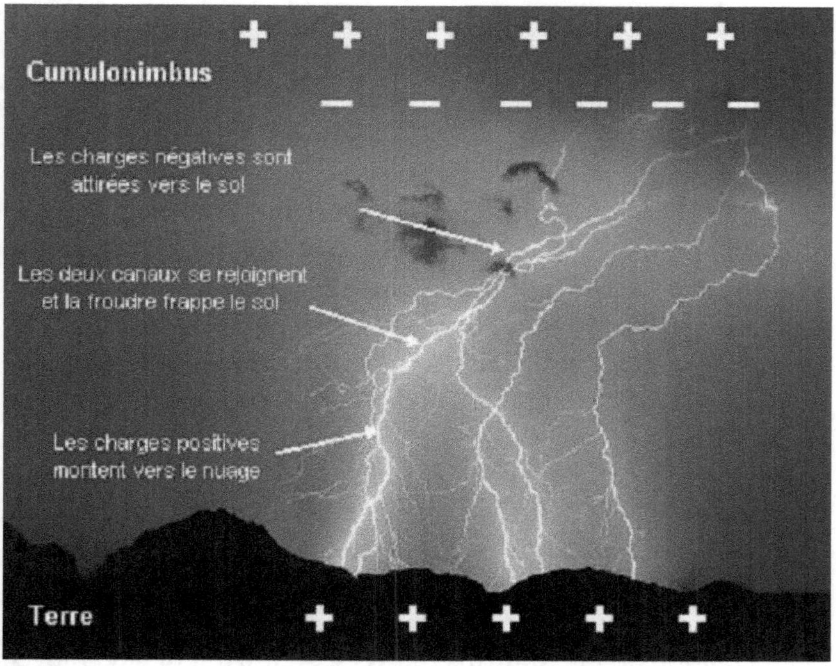

This illustration explains the formation of lightning in a thunderstorm. Credits : Astrosurf

https://trustmyscience.com/wp-content/uploads/2019/05/formation-eclair-orage.jpeg

The Universal Cosmological Model, which tends to unify and correct all Models, does not exclude explanations before analyzing them.

For the Universal Cosmological Model, the theory of the Electric Universe must be limited around the contacts of virtual particles and their effects. Since these contacts preceded the Big-bang, their authors should rather join the Universal Cosmological Model which explains the continuity of the first energies up to man by recognizing the scientific achievements of the Standard Model. All this can be understood scientifically with this analysis by the physicist J. Baker's analysis of reionisation :

"When the photons that made up the cosmological diffuse background were released, the Universe went from an electrically charged and opaque state (electrons and protons could scatter photons) to a neutral and transparent state... But the Universe we

see today is almost entirely ionized. Intergalactic space is filled with charged particles...". (Cf : J. Baker : 50 Keys to Understanding Astronomy, ed. Dunod, Paris 2016, p. 145).

The idea of the electric Universe only confirmed everything I said about the creation of electromagnetic zones generating different charges and radiation because of a non-uniform rotation speed at different levels of the Universe and I gave as a case study the study of the sun. The phenomenon is similar to what happens in terrestrial space concerning hurricanes. Indeed, the strong winds, heavy rains and swirling clouds of a hurricane produce electricity in the clouds that manifests itself as lightning, or even lightning strikes.

To support all that I have just said on this point, I recommend the scientific documentary "New Unified Field Theory : Explanation of the Dynamic Universe", which confirms that it is the Motion coming from the cause on the first virtual particles, according to the law of uncertainty, which is at the basis of all creation and it confirms the verse "...a wind of God moved over the waters" (Ge 1:2) before the Big-bang.

Therefore, we support the theories of the Electric Universe while reminding their authors that the Universal Cosmological Model includes the Big-bang.

Chapter XI
Science and Religion

Section 1
The cause of errors in classical cosmological models

1.1. The origin of energy

The main cause of the falsity of the whole classical cosmological model dealing with the origin of the Universe goes back to the origin of energy. Even the models that I have not explained here have discovered their inadequacies from everything that is developed in this Volume.

In order to talk about the origins of the Universe, we need energy for the activation of particles in the vacuum. These energies must have a source. Not believing in God as the primary Universal cause, so those who have spoken of the Pre-Big-bang Universe, be it in string theory, as in other theories, they have come to the conclusion that the Universe that preceded the Pre-Big-bang Universe can only be another Universe. Hence we speak of multidimensional Universes as in string theory or successive Universes as in Roger Penrose's CCC.

Other models also went in the same direction, because it is the explanation of the beginning that causes the problem. The beginning itself should also have an energy that should also come from somewhere.

1.2. The Principle of creation and the origin of energy

Most theories support the idea that the Universe is finite. This creates the paradox that the initial energy must come from somewhere. This leads to the assumption that there is something else beyond the Universe. Thus leads to the idea of Multiverse or successive Universes. But Rev. Dr. Sun Myung Moon's Principle of Creation is the only one that answers this question about the origin of energy and makes it so that things started with God who has the Universal prime energy in him.

Wanting to explain everything is a way of forcing me to plagiarize his book. Each researcher is free to go to a SWOT center to get more clarification on this subject and to understand the author as well.

All that we can summaries is that on the basis of the projection of the Universal primeval energy as the original energy, that everything came into existence beginning with the Universe itself.

This is confirmed even by the quote from Pope Francis that we read above, "The beginning of the world is not a result of a chaos that owes its origins to another chaos, but derives directly from a supreme Principle that creates out of love."

The Bible verse confirming this is : "In the beginning God created the heavens and the earth. The earth was formless and empty." (Ge 1:1-2).

Ge 1:1 "In the beginning God created the heavens and the earth. 2 The earth was formless and empty. "

I have already said many times that a formless and empty earth was Moses' way of saying that in the beginning the Universe was empty because physically or geographically, a formless Earth does not exist. Everything that is visible, has a form. This empty Universe according to the Bible was created by God. Astrophysicists who did not believe this created problems for themselves because they did not know at what stage to speak of the beginning. So they talk about Universes from other Universes. However, for the Bible, restored in the Universal Cosmological Model. the Universe has indeed a beginning.

Only one scientist, who believed in the Bible, pursued the origin of energy at its beginning that is to say by seeing God himself. This scientist, we have said, is the Rev. Sun Myung Moon. The fact that he discovered surprised me when I studied his teaching. That is why I had said in Volume I that whoever reads the Principle of Creation without feeling a shock will never understand the Divine Principle. Only people, who do not ask questions, will take the Divine Principle as a story and they will never feel the shock.

Moreover, several shocks created by the teaching of Divine Principles have caused some Christian conservatives to call the Rev. Sun Myung Moon a heretic. But his teaching is logical and spiritually rational. Only a verbal explanation offered by a teacher of the Divine Principle can deepen what Rev. Moon says about the origin of energy.

I even refrain from saying Rev. Sun Myung Moon's conclusions on the origins of energy in this book, because this requires a teaching that begins with the Introduction of the Principle. Even in mathematics, there are steps to be taken to solve certain equations.

Section 2
From the spiritual to the physical

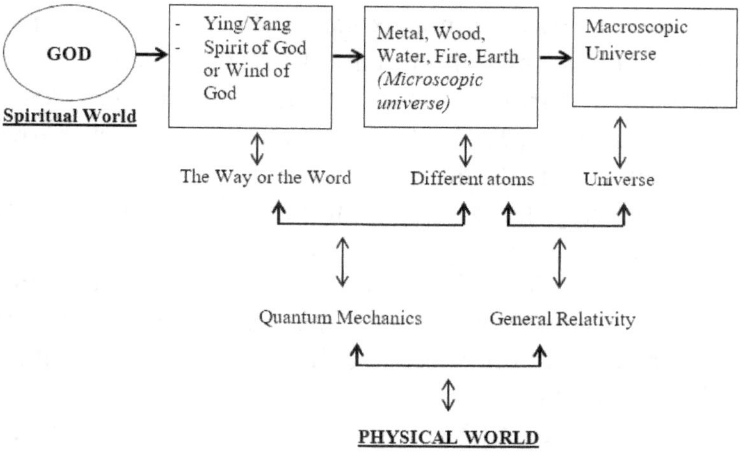

It is in the Universal Cosmological Model, which takes into account the contribution of radiation from the Pre-Big Bang to material, that the Theory of Everything can germinate, as we can read in this scientific writing "The Universe contains three sources of energy : material, radiation and quantum vacuum". (Cf : Aux Frontières de L'Univers – Du Big-bang au Quark " Marceau Felden – 2005).

In the Universal Cosmological Model this post-Big-bang scheme is also part of the Pre-Big-bang scheme. In the biblical verse "the earth was formless and empty" (Ge 1:2) ; we are dealing with a vacuum qualified as a quantum vacuum. "There was darkness on the surface of the abyss" ; here we are dealing with radiation and "the Spirit of God was moving over the waters", which symbolize energy (spirit) over material (waters).

– Radiation, at the basis of which the divine forces began to act ;

– Quantum emptiness or chaos as in the verse "the earth was formless and empty".

– Material, as concentrated energy, described in general relativity

The desire to develop this path of radiation, by piercing the quantum vacuum, until it reaches material, is one of the very complicated branches of astrophysicists. We have offered some excerpts based on their experiments to make the reader understand that everything is connected.

If radiation has been the path from divine energies to quantum energy, however, we cannot explain the behavior of particles in the latter field. Thus, the radiation between the divinity and the elementary particles being difficult to achieve, it can be understood in the experiments carried out by physicists between light, vacuum and material.

We will develop this point in Volume IV "The Enigmas Solved in the Universal Cosmological Model".

Thus, the collection "The Verses of the Scientists" extends its study of the explanation of the complicated verses of Genesis to the book of Revelation, revealing their providential fulfillment until the return of Christ.

Just as everything that astrophysicists observe is part of the effects, which are the object of their science, they must include in their discipline as complementary knowledge, the cause. And the cause is indeed the teaching of the Principle of Creation of Rev. Dr. Sun Myung Moon, as the philosophical basis of their teaching or knowledge. It is only through this teaching that they will be able to solve many riddles. This concern responds to the words of Jesus who said to these disciples :

"I speak to you of earthly things and you do not believe; how will you believe when I speak to you of heavenly things?" (Jn 3:12).

This promise of Jesus is fulfilled to this day by the Moon couple. All that remains is for the intellectual luminaries to appropriate it and bring it into their discipline.

General Conclusion

Volume I was a scientific reconstruction of the biblical scheme of creation. This restitution gave birth to a new cosmological model called the "Universal Cosmological Model". What is the place of this new model, built on the biblical foundation, in relation to other cosmological models ? Is it a copy, an extension or an original model compared to the more than twenty models of cosmology ? This is what we have tried to answer throughout this Volume II.

We have said that it is to spare the word Bible from any judgment, so I have baptized my scientific restitution of the biblical scheme of creation under the name of "Universal Cosmological Model". This model will henceforth serve as the basis for all arguments and models that will support the biblical scheme of creation.

Thus, in this Volume II, we have offered our analysis of cosmological notions according to the vision of the Universal Cosmological Model. We have used the observations of astrophysicists for this purpose, without, however, binding ourselves to their analyses. We approved what we found right and we rejected, not the phenomena, but certain interpretations related to them to atheism. We have also contributed by trying to solve certain enigmas, or by offering the scheme for their solution.

Moreover, starting with the description of the origin of the Universe, through the Big Bang to humankind, the Universal Cosmological Model, encompasses and corrects all other cosmological models, even those I have not mentioned in this book.

From now on, the other models that will be created after 2020 will no longer make the big bang a singularity. With the Universal Cosmological Model, all of humanity is now convinced that the Big Bang is not just a theory, but a physical reality described biblically as "the separation of the waters from the waters" (Gn1:6) and in the Qur'an as «Do not the unbelievers see that the heavens and the earth were originally a compact mass? We then split them" (Qur'an 21:30).

Compact mass is what the Bible called the waters before its separation and science called the universe condensed before the big bang. This is why the Universal Cosmological Model is universal.

My words are not sacred messages. Errors can be detected and once brought to my attention, they can be corrected in some versions.

So thank you for the understanding and contribution you want to share with us.

Flavien PHANZU MWAKA

Bibliography

I. SCIENTIFIC LITERATURE

– Aurélien Barrau, Big-bang et au-delà, balade en cosmologie, ed.Dunod, Paris

– Aux Frontières de L'Univers – From the Big-bang to the Quark "Marceau Felden – 2005

– Basset, Bruce & Edney Ralph, *Relativité*, Icon books ltd, Londres, 2009.

– Bender, Lionel : *les plantes*, éd. Artis-Historia, Bruxelles, 1989.

– Douce Roland & Postaire Eric, *Les origines de la vie*, Gallimard, 2016.

– Galfard, Christophe, *E=mc*, Flammarion, 2017

– Heidmann, jean : Sommes-nous seuls dans l'univers, Fayard, Paris, 2000.

– Herrmann, Joachim, *Atlas d'Astronomie*, ed. Stock, Paris, 1976

– Igor & Bogdanov Grichka, *Avant le Big-bang*, Grasset, 2004.

– *J.Baker, 50 clés pour comprendre l'astronomie, éd. Dunod, Paris 2016*

– Jean Bricmon, La mécanique quantique pour non physiciens, B-1348 Louvain-la-Neuve, Belgium

– Klein, Etienne *: Les secrets de la matière*, éd. J'ai lu, Paris, 2018.

– Klein, Etienne : Discours sur l'origine de l'univers, éd. Flammarion 2016.

– Le Principe divin, niveau 4, Paris

– Pluchet, Blandine, *La physique quantique pour les nuls*, éd. First, 2018.

– Reeves, Hubert : *L'heure de l'ivresse; L'univers a-t-il un sens*? Le Seuil, 1986

– Sagan, Carl : *Cosmic Connection*, ed. du Seuil, Paris, 1975

– Sarrailh, Claude, *Le secret de la vie*, éd. Librinova, 2018

– *Stephen Hawking, Une brève histoire du temps, trad.I. Naddeo-Souriau, Paris, Flammarion, 1989*

– Whyntie & Pugh, Oliver, *La physique des particules en image*, EDP Sciences, 2014.

II. RELIGIOUS BOOKS

– Louis Segond Bible
– Jerusalem Bible
– BABYLON THE GREAT IS FALLEN– the kingdom of God has begun its reign, Watchtower, JW.org,
– Presentation of the Divine Principle : Culture and Peace, Paris, 2013
– The Divine Principle, Level 4, Introduction
– FFPMU : Unification Thought
– UNIVERSAL LIFE : The fields of radiation, Würzburg 1, 1986.

III. COLLECTIONS

– Les défis de la science : *l'origine de la vie*, édité par RBA coleccionables, S.A.U. 2018, Barcelone.
– Les défis de la science : *La théorie du Tout*, édité par RBA coleccionables, S.A.U. 2019, Barcelone.
– Les défis de la science : *Au-delà du Boson de Higgs*, édité par RBA coleccionables, S.A.U. 2018, Barcelone.
– Voyage dans le cosmos : *Les trous noirs*, édité par RBA coleccionables, S.A.U. 2020, Barcelone.
– Voyage dans le cosmos : *Les exoplanètes*, édité par RBA coleccionables, S.A.U. 2020, Barcelone.
– Voyage dans le cosmos : *Le boson de Higgs*, édité par RBA coleccionables, S.A.U. 2020, Barcelone.
– Voyage dans le cosmos : *L'espace-temps quantique*, édité par RBA coleccionables, S.A.U. 2020, Barcelone.
– Voyage dans le cosmos : *Les univers parallèles*, édité par RBA coleccionables, S.A.U. 2020, Barcelone.
– Voyage dans le cosmos : *Les trous noirs*, édité par RBA coleccionables, S.A.U. 2020, Barcelone.
– Voyage dans le cosmos : *La matière noire*, édité par RBA coleccionables, S.A.U. 2020, Barcelone.
– Voyage dans le cosmos : *Le vide et le néant*, édité par RBA coleccionables, S.A.U. 2020, Barcelone.

IV. COURSES & PRACTICAL WORK

- Lachieze-Rey, Marc : *Initiation à la comologie*, 5è éd, Dunot, Paris, 2013.
- Matthews, P.T., *Introduction à la mécanique quantique*, Dunod, Paris, 1970.
- Ir. MPANZU KIADI BABAKA et l'Ir. NIMI M., Travaux pratiques de mécanique quantique : explication de la mécanique quantique, UPN –RDC– 2017, inédit

V. MAGAZINES AND NEWSPAPERS

- Les secrets de l'espace 07 : # HORS SERIE#, Les mystères de l'Univers
- C. Bonneau, Science&Vie Hors-séries n° 288, 09/2019, p.55
- Hugo Leroux, Science&Vie Hors-Série n°288, 09/2019, p.58
- La Libre Belgique 2005. Guy Duplat
- Des signaux extraterrestres détectés par les équipes deStephen Hawking; Paris Match| Publié le 04/09/2017 par Camille Hazard

VI. ENCYCLOPEDIAS

- De l' Astronomie de l'Espace : Les forces fondamentales, Hachette, 2019.
- De l' Astronomie de l'Espace : Les trous noirs, Hachette, 2018
- De l' Astronomie de l'Espace : Matière noire et énergie sombre, Hachette, 2019.
- Edey, Maitland A., Les origins de l'Homme; Le Chaînon manquant, éd. France Loisirs, Paris, 1984
- Geographic exploration. "Microsoft® Encarta® 2009 [DVD]. Microsoft Corporation, 2008
- Grint, Damian, *Les mystères de l'univers*, éd. Christophe Colomb, Paris, 1985
- Joe Bergeron, Encarta ® 2009 Modeling a black hole
- Gamlin, Linda, *Les origines de la vie*, éd. Artis-Historia, Bruxelles. 1989
- Melvin & G. Berger : *Qu'est-ce qui fait onduler l'océan*? ed. Nouveaux Horizons– ARS, Paris 2012
- Moïse. " Microsoft® Encarta® 2009 [DVD]. Microsoft Corporation, 2008
- P. Batchelor : *L'énergie, le son et la lumière*, éd. Nouveaux Horizons, ARS, Paris, 2016

– Zoppi, Moninga, Les Merveilles de l'univers, Médiaspaul, Kinshasa, 2009.
– "conservation, lois de." Microsoft® Encarta® 2009 [DVD]. Corporation, 2008
– "dinosaures." Microsoft® Encarta® 2009 [DVD]. Microsoft Corporation, 2008
– Électron : Encarta
– "gravitation." Microsoft® Encarta® 2009 [DVD]. Microsoft Corporation, 2008
– "Olbers, paradoxe d'." Microsoft® Encarta® 2009 [DVD]. Microsoft Corporation, 2008
– "Pangée." Microsoft® Encarta® 2009 [DVD]. Microsoft Corporation, 2008
– Gravitationnelles, ondes." Microsoft® Encarta® 2009 [DVD]. Microsoft Corporation, 2008
– THÉORIE DU BIG-BANG : encarta
– Wikipédia; *Particles élémentaires*
– Wikipedia; Metaphysical
– Bételgeuse : Wikipédia
– *Théorie des cordes : Wikipédia*
– Wikiedia; Modèle Standard
– Wikipédia : BIG-BANG
– Wikipédia : Brane
– Wikipédia : Empédocle
– Wikipédia : GN-z11
– Wikipédia : Histoire de l'Univers
– Wikipédia : L'énergie d'activation
– Wikipédia : Les particules subatomiques
– Wikipédia : supersymétrie
– Wikipédia; BRANE
– Wikipédia, Cellules
– Wikipédia, Multivers
– Wikipédia.fr, galaxies
– wikipedia.org/wiki/Univers
– zevengeur.wordpress.com/2014/01/16/histoire-de-lUniverspartie -4-le-modele-standard-de-la-physique
– zevengeur.wordpress.com/ la-chromo-dynamique-quantique

– *Zevengeur, www.blogs, zevengeur, Modèle Janus, le futur de la cosmologie, le12/01/ 2018*

VII. WEB SITES

– astropolis.fr/Le Big-bang : Instant zéro ou "singularité initiale
– astrofiles.net/astronomie-nicolas-copernic
–astrophysique/hubble-bubble-notre-galaxie-est-elleenfermee-dans-une-bulle-les-explications-de-david-elbaz Par Olivier Lascar, le 03.01.2017
– Blogs zevenzeur; Introduction à la physique fondamentale, publié le 18/02/2013
– *boowiki.info/art/cosmologie/cosmologie-conforme-cyclique*
– Express.fr, les premières secousses du Big-bang expliquées par Grichka Bogdanov, par Jérémie Pham-Lê, publié le 18/03/2014
–fredzone.org/stephen-hawking-pense-toujours-que-nousdevons -nous-mefier-des-extraterrestres-554
futurasciences.com/buildsv6/images/mediumoriginal/d/e/7/de7da23f 61_65821Univers-jean-pierreluminet.jp
– futurascience.com/10 questions essentielles sur l'Univers/ le 19/08/2019
– futura-sciences.com/sciences/definitions/physiquephysique-quantique-13197
– futura-sciences.com/sciences/definitions/Univers-voielactee-3729 par Floriane Boyer
– *Futura-sciences.fr*, La bizarre asymétrie du rayonnement fossile : une énigme résolue ? le 02/01/2009 par Laurent Sacco).
– futurasciences.com ; Laurent Sacco ; le 02/03/2018
– futura-sciences.com/sciences/définitions/chimie-zircon-4668
– Futurasciences, JP Luminet, Singularités et temps zéro
– *futura-sciences.com /Luigi Bianchi, Nous serions perdus dans un Univers isotrope*
– *futura-sciences.com/sciences/actualites/cosmologie-preuveavant -big-bang-roger-penrose*
– futurascience.com/10questions essentielles sur l'Univers/ le 19/08/2019
– futurascience.com, Hayabusa 2 : Mascot s'est posé sur Ryugu et commence ses activités scientifiques, par Remy Decourt,Publié le 04/10/2018).

– futura science.com, Origine de la vie : nouvelle piste avec des poussières interstellaires, *par Laurent Sacco, le 23/11/2017*
– gurumed.org/2019/03/21/ la-sonde-japonaise-hayabusa-2-montre-que-lastrode-ryugu-est-tonnamment-sec
– Les grands Entretiens d'Actu-Philosophia,Entretien avec Aurélien Barrau, 5 juin 2012, par Florian Forestier sur web
– matierevolution.fr/spip.php ? auteurs2Matière noire, énergie noire : le chaînon manquant ? le 13/11/2007, par FaberSperber, Robert Paris
– matierevolution.fr/spip.php ?article38Le vide,... pas si vide, le23/06/2010, par Robert Paris
– *matierevolution.*fr/spip.php ?article, Les bulles de vide et la matière, le 16/05/2008, Robert Paris
– *matierevolution.fr/spip.php ?article, Big bang ou pas Big-Bang, le 15/05/2009 par FaberSperber, Robert Paris*
– *motherboard.vice.com/a-la-rencontre-des-partisans-de-latheorie -de-l-Univers-electrique par Sarah Scolesfévr. 22 2016*
–sciencesetavenir.fr/espace/asttrustmyscience.com/expansion-Univers-acceleration-trop-rapide-pouretre-expliquee-par-la-physique-actuelle, par Thomas Boisson, le 26/02/2018
– sciencesetavenir.fr/la-preuve-de-l-inflation-de-l-Univers-n'a-toujours-pas-ete-observee ; par AzarKhalatbari, le 02.02.2015
– *sciencetonante.wordpress.com/2013/09/30/les-7-merveillesde-la-mécanique-quantique*
– tdg.ch/preuve-inflation-cosmique-big-bang-faite
– Techno-Science.net. Par Bongo, le 23/03/2014
– trustmyscience.com/ nouvelle-etude-matiere-noireplus-ancienne-que-big-bang/ par Stéphanie Schmidt, 9 août 2019
– trustmyscience.com/voici-a-quoi-ressemblerait-laterre-si-toute-la-glace-venait-a-fondre par Julien Claudet, le 17 juillet 2017
– trustmyscience.com/technologie-produit-oxygene-depuisc o2-pourrait-aider-colonisation-mars ; par Thomas Boisson, le 28/05/2019
– trustmyscience.com/hadron/ par Thomas Boisson, le 3 août 2018
– *trustmyscience.com/*l'inflation cosmique recréée en laboratoire grâce à un condensat de Bose-Einstein, Thomas Boisson, 2/05/2018

– *vinsvignesvignerons.com/Geologie/Une-breve-histoire-dela-planete-terre*

VIII. Scientific documentaries

– « Au-delà du Big-bang »
– « L'Univers de STEPHEN Hawking »
– « Le mystère de l'Univers : La lumière et les ténèbres ».
– « Les secrets du Cosmos : la théorie du champ de Higgs
– « Nouvelle Théorie du Champ Unifié (Explication de la Gravitation)
– « *Nouvelle théorie du champ unifié : explication de l'Univers Dynamique* »
– « Physique quantique pour les nuls
– « Quelle est l'origine de l'eau sur la terre »
– « Toutes les planètes du système solaire »
– Conférence : Qu'y avait-il avant le Big-bang
– A.E. Wilder Smith : « les origines de l'Univers »
– Alice Harper et Nigel Henbest : « les origines de la terre »
– Colin Campbell : « Les mystères de l'Univers – au-delà du froid »
– D. Louapre, « les 7 merveilles de la mécanique quantique »
– David Louapre, « la mécanique quantique en 7 idées »
– Garry Parker : « Big-bang »
– Hubert Reeves Conférence-Documentaire « que pensez-vous de la déclaration de Stephen Hawking ? »
– Hubert Reeves ; « D'où provient l'énergie du Big-bang »
– Iolande Cadrin-Rossignol : « Du Big-bang au Vivant »
– J.P. Luminet, conférence-documentaire « la forme de l'espace »
– Jean Pierre Martin, conférence-documentaire « Et la lumière fut : ou l'histoire du Big-bang »
– Laura Verkland Armstrong ; « L'Univers – Les Mystères du Big-bang »
– Gary Wright : Le Soleil, Source De Vie Et De Danger
– Matt Hickey : « Cosmos – Au-delà du Big-bang »
– Pascal Picq : conférence-documentaire "Premier Homme"

Sommaire

Dépôt légal février 2022

Librinova

www.librinova.com

www.ingramcontent.com/pod-product-compliance
Lightning Source LLC
LaVergne TN
LVHW050807061126
851857LV00012B/716
* 9 7 9 1 0 2 6 2 7 5 1 2 1 *